HOTEL MANAGEMENT AND OPERATIONS

Hotel Management and Operations

Editor

Delphina Fonseca

scitus
academics

Hotel Management and Operations
Edited by **Delphina Fonseca**

ISBN: 978-1-68117-545-4
Library of Congress Control Number: 2016942949

SCITUS Academics LLC,
www.scitusacademics.com
Box No. 4766, 616 Corporate Way,
Suite 2, Valley Cottage,
NY 10989

Notice

Printed in United States of America on Acid Free Paper

PREFACE

Hotel Management is an area of study that covers a wide range of topics concerned with the operational aspects of hoteliery. Hotel operations are varied and cover topics as wide ranging as marketing, eco-tourism, leisure, business administration and management. It is the job of the Hotel Manager to coordinate the many operational tasks of running a successful hotel. Managing a hotel requires knowledge and skills within a wide variety of fields. A hotel manager, hotelier, or lodging manager is a person who manages the operation of a hotel, motel, resort, or other lodging-related establishment. Management of a hotel operation contains, but is not limited to management of hotel staff, business management, upkeep and sanitary standards of hotel facilities, guest satisfaction and customer service, marketing management, sales management, revenue management, financial accounting, purchasing, and other functions. The size and complexity of a hotel management organizational structure varies significantly depending on the size, features, and function of the hotel or resort. Administrative functions for a small-scale hotel such as Accounting, Payroll, and Human Resources may normally be handled by a centralized corporate office or solely by the Hotel Manager.

Hotel Management and Operations presents a variety of viewpoints on the duties, responsibilities, problems, and opportunities encountered there. The book continues to provide a comprehensive and lucid coverage of the subject. The text gives a practical approach, challenge readers to recognize the central issues involved in complex management problems, understand the structure and resources of the department in question, and discover solutions that may help in supervision other hotel resources and departments.

TABLE OF CONTENTS

7

A STUDY ON THE INFLUENCE OF TECHNOLOGIES APPLIED TO ORGANIZATIONAL PERFORMANCE LEARNING IN THE HOTEL SECTOR

Pedroche MSC[1*], Antón JMR[1], Andrada R[2] and Karaboytcheva MK[3]

[1]Department of Business Organisation, Autonomous University of Madrid, Carretera de Colmenar Viejo, km. 15, 28049, Madrid, Spain

[2]Department of Applied Economics, Autonomous University of Madrid, Carretera de Colmenar Viejo, km.15, 28049, Madrid, Spain

[3]Department of Quantitative Methods and Economic Theory, University of Alicante, Carretera San Vicente del Raspeig S/N, 03690, Alicante, Spain

ABSTRACT

Hotel sector is one of the most influenced by the new information and communication technologies (ICTs) since it generates a vast amount of information which circulates through these new technologies. Additionally, the latter could be a potential source of organizational performance. This paper has an exploratory character and analyses how the new ICTs influence the hotels' organizational performance. Thus, a descriptive statistical analysis and a study of statistical associations through Cramer's V have been carried out using data from 106 Valencian Community hotel establishments. The results show, on the one hand, that Internet and Intranet are the most influential technologies on intra-organizational level and, on the other hand, show that the Central Reservation Systems -CRS- and web 2.0 are the most important on an inter-organizational level. Both technologies affect most of the analysed results at organizational level.

KEYWORDS

Organizational learning; ICTs; Organizational performance; Hotel sector

1. INTRODUCTION

The Information and Communications Technology (ICT) development has altered the ways in which tourism and hotel sector companies operate [1], driving significant changes and providing new opportunities for the industry [2]. Technological advances have been linked to organizational learning [3-5], which can be understood as the organizations' ability to maintain or improve their performance based on experience [6]. This ability involves acquiring, sharing and using the knowledge. Therefore, this is a process which increases the knowledge created in an organized and transformed way within the organization's knowledge system [6]; and it influences, according to Petrash [7], Gupta and Govindarajan [8] and Olivera [9], organizational performance in the tourism sector.

The technologies have positively contributed to the hotel industry, which has led the hotels to invest substantially in them [10] in order to improve both, internal processes and the service provided to consumers [11]. The research in this field has shown that the initiative to use ICTs can lead to achieve a sustainable competitive advantage [12,13], although in a recent study, Mihalič and Buhalis [14] show that ICTs do not directly increase the productivity of companies.

Since the technology plays an important role in the operations and management of the hotel, both managers and researchers must be informed about its development [1], so it becomes necessary that the researches deepen the ICT's general impact on the hotel performance [14,15]. In this connection, our paper aims to analyse the utility level of different ICTs within the context of organizational learning granted by the managers of 3, 4 and 5 star hotels in Valencia, but especially how the ICTs influence organizational results.

To achieve this goal, first a review of the related literature is done; secondly, the methodology used is explained in details, and then a descriptive statistical analysis of the data is carried out in order to find what technology and what organizational performance are most valued by the respondents in context of learning. Subsequently, association between ICTs and organizational performance is analysed and, finally the achieved results are presented and discussed to reach the conclusions.

2. THEORETICAL FRAMEWORK

2.1 Information and communications technology

Up to now, the literature has been more focused on the introduction and development of new ICT's, as well as on the costs and benefits involving its implementation, than on the impact of sharing knowledge and its use [1,16-23]. Therefore, this study pretends to shed more light on the latter. Technological progress and the tourism sector have been linked for decades [5]. The hotel industry has begun to use them into its management processes since the seventies of the last century. The Property Management Systems (PMS) improving interoperability and interconnectivity [24], as well as the Central Reservation System (CRS), providing automatic management of reserves [25] began to be used in hotels in that decade. The Global Distribution Systems (GDS), systems in which take part a group of tourism suppliers that offer information both airlines and hotels, while allowing to make bookings appeared in the late eighties. The Internet revolution, known as network of networks [26] came in the nineties, but it is since 2000 when a true transformation in the ICT's have been taking place with the appearance of other technologies, like Wireless-radio connection and Wi-Fi, thank to which Internet can be accessed anywhere without requiring a physical connection. At that time also appeared the blogs as a platform through which users collaborate in the construction of knowledge. In the case of tourism, we can mention Web 2.0 or Travel 2.0 which link the concepts of virtual communities or social networks and, according to Xiang and Gretzel [27], apply them in the search of travel information related. Finally, we can add the proliferation of different mobile technologies such as Personal Digital Assistants (PDA), 3G, Global Position Systems (GPS), etc.

When applied to organizational learning, ICT's include a variety of ways to create, capture, manipulate, communicate, share, present and use information [28]. In this study, the main ICT's have been grouped into two blocks: (1) for internal use among hotel employees and (2) oriented to relationships with stakeholders [10], using as a measuring unit the ICT's utility level in the process of organizational learning [15].

2.2 ICT and organizational learning

Fiol and Lyles [29] defined organizational learning as the process of improving actions through better knowledge and understanding. This organizational learning is supported by the management of knowledge creation in the organization and ICTs are a mean that can facilitate knowledge sharing and application [30]. In this line, some authors like Robey et al. [3], Moreno and Vargas [31], Martín-Rojas et al. [13] state that ICT's can facilitate the learning process, i.e., the processes of creating, accessing, transferring, sharing and codification of knowledge [31,32] and information [33].

One of the hotel sector's main features is that their services processes are becoming knowledge-based priority due to the huge influence of the ICT's use [34]. In this sense, Figueroa [25] argues that ICTs are means to improve the management of information and to create new tourism products and services. Companies operating in the hotel sector have to become more innovative and increase their technological capacities in order to improve their organizational learning [35,36].

Different types of technologies have been adopted in the hotel industry during the past two decades) [37]. Thus, Buhalis [26] distinguished those technologies in the sector oriented to intraorganizational functions which exist within one organization from those oriented to inter-organizational functions-which occur between different tourism companies and, finally, distinguish those which are customer-oriented. The same author points out as examples of the technologies used at intra-level - Internet, Intranet and PMS. The email cannot be ignored at this level, as it allows not only an interaction with external agents, but also between the organization's members [38]. Furthermore, at inter-organizational level, Buhalis [26] mentions the CRS, email and website; and at consumers relations level, repeats the Internet, the website, email and the CRS. It should be added Web 2.0, which has also had a big influence on the tourism sector [37] and the CRM because its technology allows the hotels to collect the preferences of customers and learn how to use them in the future [39]. In consequence, we consider in this research all these technologies because of their close relationship with the organizational learning.

2.3 Results of the use of ICTs as a learning tool

Certainly many hotel executives evaluate the technologies' potential to achieve significant advantages in operational and strategic direction, so the hotel organizations have relied on technologies to take decisions and to develop their business [40]. In order to improve organizational performance, ICTs are becoming a strategic asset for the tourism and hotels business [41]. Since the rapid development and commercialization of the ICTs, there are many advantages for hotels and related industries which are incited to adopt these technologies [42]. Some researches show that investment in ICTs could improve the performance and productivity of a hotel [40,43,44], but others suggest the opposite, it is to say, the investment does not improves return [45,46]. The present study attempts to follow this research line, enquiring whether an extraordinary organizational performance can be achieved through the ICTs.

In relation to the returns that tourism sector organizations could achieve through ICTs, we observe in the literature that: (1) if ICTs are properly directed, they would help tourism companies to achieve a competitive advantage to maintain price leadership in the market or to differentiate their products and services [24], which can be assimilated with organizational performance as a strategic renewal or new product development and innovation; (2) investment in ICTs in the hotel sector benefits the hotels in the way that it allows customers to have a better experience and the hotel staff work more efficiently to better assist customers [11] which can increase loyalty; (3) the use of ICTs can bring knowledge and information in the focus of the hotel organization's competitive profile [47], which can improve its competitive position. Law and Jogaratnam [11] point out that for hotel professionals is important to introduce proactively ICTs in their hotels in order to improve service quality; (5) There are many more benefits that can be obtained through the adoption of ICTs, for example, reducing operational costs, enhancing customer satisfaction, increasing market share and increasing operational efficiency [26,46,48,49] and, therefore, they are also linked to the improve of profits, sales or targets.

3. METHODOLOGY

This quantitative study is focused on the hotel sector in the Valencian Community, as it is one of the main tourist destinations in Spain. The Valencian Community was the sixth community in number of tourists' visits, with a total of 7.293.995 travellers in 2014, mainly for sun and beach tourism, and represent 8% of all international and national travellers in Spain. Additionally, the hotel sector in Valencia is one of the industries with major importance. That Community has 960 hotel establishments with a total of 13.847 people employed in the sector [50].

To achieve the proposed research aim, a questionnaire based on the most relevant literature was designed. It consisted of 32 questions related to a) the use of ICTs in the field of the internal organizational process; b) the relationships maintenance with other stakeholders linked to the analysed hotels; c) the organizational results reached by the analysed in the survey hotels. The survey also included questions in order to categorize respondents and the hotels where they work. Table 1 shows the technical specifications of the research.

To contrast the validity of the measurement instrument, a panel composed by four academic experts on the topic was set up with the aim to use their theoretical and empirical experience to analyse and evaluate the full list of items that were collected from the literature. Thanks to their points of view after setting the first survey, we went on to perform a pre-test which was evaluated by three senior managers from 3, 4 and 5 star hotels respectively. The final questionnaire was established considering their remarks.

According to the results, most of the 106 hotels that were sampled belonged to chains: 58.2%. 70.2% of them are owned by national chains; 64.7% are directly operated by the owners; most of the clients who were staying there did so for leisure: 57.7%; higher percentage of hotels 36.3% have been between 20 and 50 years in the industry; 26.2% have between 100 and 199 rooms and 54.4% of the hotels studied have between 10 and 49 employees being, and therefore classified as small businesses by the European Commission [51].

Table 1: Technical details.

Population or Universe	395 hotels of 3, 4 and 5 stars in the Valencian Community
Sample size	106 hotel establishments
Analysis units	Hotel establishments
Response rate	26.84%
Sampling error	8.1%
Confidence level	95%
Fieldwork period	November 2013-April 2014
Tracking	Phone
Survey's delivery method	Post, personal, email
Respondents	Senior managers or Department heads
Source: Compiled by authors	

4. RESULTS

Once data collection process was completed, first, a descriptive statistical analysis was performed to, then, conduct a study of existing statistical associations of each variable contained in ICTs within the organizational learning in the organization and with other stakeholders, as well as with different variables which shape the organizational results achieved by the hotels.

The first results refer to the use of the ICTs by the hotels studied here. Table 2 shows how the most used technology internally by the employees is Internet, with a mean value of 4.85, having the lowest standard deviation 0.494. Internet is the most used by the employees' technology in their deals with customers, suppliers, etc. It is the most valued by respondents with a mean of 4.75 and a low standard deviation -0.61.

In reference to organizational results, it can be seen in Table 2, that increased customer loyalty is the most valuated result, with a mean of 4.10 and the lowest standard deviation, 0.804. It is particularly important that the third most appreciated variable is the quality improvement, with a mean value of 4.04 which is closely linked with increased loyalty.

Table 2: Descriptive statistical analysis of the variables.

	Tecnologies	Mean	Std. dev.
Intra	Internet connection	4.85	0.494
	Email	4.46	0.978
	Existence of Intranet	4.43	1.060
	PMS –Property Management System-	4.40	1.097
Inter	Internet connection	4.75	0.619
	Email	4.73	0.595
	Web site	4.61	0.756
	Web 2.0/Blogs	4.39	0.938
	CRS -Central Reservation System-	4.35	0.983
Organisational Performance	Increased customers loyalty	4.10	0.804
	Competitive position improvement	4.06	0.807
	Quality improvement	4.04	0.827
	Efficiency increase	3.98	0.894
	Costs reduction	3.94	0.893
	Sales growth	3.88	0.885
	Fulfilment of objectives	3.87	0.906
	Increase of profits	3.72	0.881
	Strategic renewal	3.69	0.909
	Innovation	3.69	0.970
	Development of new products	3.63	0.949
Source: Compiled by authors			

Then, analysis of the existing statistical association between each type of technology used, both internally and externally, and organizational results achieved by the hotels was carried out. To do so, we used the Cramer's V which is a correction that can be applied to Pearson Chi-Square. This allows obtaining a maximum index value of 1 (indicating the largest association between quantitative or qualitative variables) and a minimum value of 0 (which represents no association).

In addition, to carry out the contrast of the null hypothesis of no association through Cramer's V, the critical value of probability is tested. Usually, if it is less than or equal to 0.05 said hypothesis is rejected and therefore a statistically significant association between each pair of variables considered exists.

Table 3. Internet connection- Intra-organizational level - Organizational results.

Internet connection – Performance		**Value**	**Approx. Significance**
Internet connection - Market share increase	Cramer's V	0.367	0.000
Internet connection - Increase of profits	Cramer's V	0.273	0.022
Internet connection - Strategic renewal	Cramer's V	0.593	0.000
Internet connection - Efficiency increase	Cramer's V	0.167	0.711
Internet connection-Sales growth	Cramer's V	0.320	0.005
Internet - Fulfilment of objectives	Cramer's V	0.325	0.000
Internet connection - Development of new services	Cramer's V	0.317	0.001
Internet connection - Customers loyalty increase	Cramer's V	0.260	0.011
Internet connection - Innovation	Cramer's V	0.281	0.003
Internet connection - Quality improvement	Cramer's V	0.246	0.023
Internet connection - Competitive position improvement	Cramer's V	0.262	0.039
Internet connection - Costs reduction	Cramer's V	0.204	0.352
Source: Compiled by authors			

Regarding the results on technologies use at intra-organisational level - Internet, Intranet, PMS and email, it can be seen that:

- According to the results, internet connection is one of the tools that are most related to a big part of the organizational results analysed. One exception is the operational efficiency increase and the costs reduction (Table 3) although some authors like Wan [52] argue that the use of Internet does not give a guarantee of success or a competitive advantage.

- Regarding Intranet, there was also a significant association with several organizational results analysed, but in a lesser degree than that with Internet. These results are related mainly to financial and customers service improvement (increase of profits, strategic renewal, sales growth, achievement of objectives, increased customers' loyalty, quality improvement and costs reduction)

(Table 4). In any case, Intranet is a tool that can benefit internally every member of the hotel as it allows conducting communications and interactions [53], which can improve competitiveness. Thus, Intranet, along with Internet, as proved in this study, in line with Garbin [53], is considered as one of the technologies with greatest impact on tourism sector.

- The email had a lower association with organizational results at the hotels employees' level of use. It only related to the strategic renewal, fulfilment of objectives and customers loyalty increase, results linked to the employees' work (Table 5). On the contrary, according to the literature, the email is more related to knowledge in organizations development [54].

Table 4, Intranet connection- Intra-organizational level - Organizational results.

Internet connection – Performance		**Value**	**Approx. Significance**
Intranet connection - Market share increase	Cramer's V	0.233	0.111
Intranet connection -Increase of profits	Cramer's V	0.297	0.002
Intranet connection - Strategic renewal	Cramer's V	0.337	0.000
Intranet connection - Efficiency increase	Cramer's V	0.226	0.153
Intranet connection - Sales growth	Cramer's V	0.317	0.002
Intranet - Fulfilment of objectives	Cramer's V	0.287	0.010
Intranet connection - Development of new services	Cramer's V	0.248	0.054
Intranet connection - Customers loyalty increase	Cramer's V	0.280	0.015
Intranet connection - Innovation	Cramer's V	0.252	0.064
Intranet connection - Quality improvement	Cramer's V	0.270	0.026
Intranet connection - Competitive position improvement	Cramer's V	0.231	0.123
Intranet connection - Costs reduction	Cramer's V	0.300	0.001
Source: Compiled by authors			

Table 5. Intra-organizational email - Organizational results.

Email-Performance		Value	Approx. Significance
Email - Market share increase	Cramer's V	0.227	0.149
Email - Increase of profits	Cramer's V	0.222	0.180
Email - Strategic renewal	Cramer's V	0.293	0.003
Email - Efficiency increase	Cramer's V	0.220	0.199
Email - Sales growth	Cramer's V	0.259	0.097
Email - Fulfilment of objectives	Cramer's V	0.471	0.000
Email - Development of new services	Cramer's V	0.237	0.092
Email - Customers loyalty increase	Cramer's V	0.366	0.000
Email - Innovation	Cramer's V	0.229	0.163
Email - Quality improvement	Cramer's V	0.223	0.200
Email - Competitive position improvement	Cramer's V	0.219	0.208
Email - Costs reduction	Cramer's V	0.201	0.380
Source: Compiled by authors			

- The Property Management Systems were more correlated to organizational results than the email. Thus, they are linked to strategic renewal, customer loyalty increase, quality improvement and cost reduction (Table 6). These results are linked to the very purpose of this tool, which is to improve the management of both the back-office, that includes, among others, the financial and commercial arrangements, and the front office, which would be linked to other related to customers activities: reception, telephone connections. etc. [27]. Additionally, the study of these authors shows that hotels using PMS reach higher productivity rates than those which do not use it. Finally, we analysed organizational results achieved related to the use of technologies at inter-organizational level: Internet, email, website, CRM, CRS, Web 2.0 and blogs. We have reached the following conclusions:

- The use of Internet with the stakeholders in the analysed hotels does not imply so good organizational results as those achieved by its use among employees. They include strategic renewal, sales growth, fulfilment of objectives and, finally, one which most sense has for the remarkable link between Internet and customers: customers loyalty increase (Table 7). Internet itself can be considered as an important innovation that can contribute to the increase of companies' competitiveness, as it facilitates the relationship with customers [55]. Then, Internet is a new communication and distribution channel for consumers and suppliers of products and services in the tourism sector [56]. Internet has been highlighted as one of the most influential technologies commercially to the consumers' behaviour [34,57,58].

- The e-mail, which is conceived as a tool for inter-organizational learning, almost did not have relationship with organizational results of the analysed hotels. Thus, it only had a significant association with the strategic renewal and fulfilment of objectives (Table 8). The email has been seen as a mean of marketing with commercial purposes and results, but lately its use has been limited by the appearance of spam [59].

- The website had a greater effect on organizational performance because it leads to increase in the efficiency, fulfilment of objectives, customer loyalty increase. It is a key tool in the hotel sector's interaction with customers and the possible increase in consumer buying int ention depends on its quality [60]. Finally, it leads to innovation (Table 9). The websites are a key tool to attract customers, especially through its design [37,61].

- Customer Relationship Management (CRM) is a technology based on deep insight into our customers' identification in order to adapt products and services to their tastes and preferences. It was associated with a considerable number of organizational performances as market share increase, profits increase, strategic renewal, development of new services, quality improvement and costs reduction (Table 10). The literature, in general, is based on the fact that the CRM increases customers' satisfaction and, therefore, leads to customers' retention [62]. Notwithstanding, we have to consider that the research that reveals the relationship between CRM and the performance of a hotel is scarce [62] and that the existing studies reject its impact on organizational performance [63,64].

Table 6 PMS at intra-organizational level – Organizational results.

PMS-Performance		Value	Approx. Significance
PMS - Market share increase	Cramer's V	0.217	0.455
PMS - Increase of profits	Cramer's V	0.223	0.396
PMS - Strategic renewal	Cramer's V	0.321	0.002
PMS - Efficiency increase	Cramer's V	0.247	0.171
PMS - Sales growth	Cramer's V	0.255	0.098
PMS - Fulfilment of objectives	Cramer's V	0.228	0.350
PMS - Development of new services	Cramer's V	0.260	0.093
PMS - Customers loyalty increase	Cramer's V	0.325	0.004
PMS - Innovation	Cramer's V	0.251	0.172
PMS - Quality improvement	Cramer's V	0.281	0.049
PMS - Competitive position improvement	Cramer's V	0.210	0.543
PMS - Costs reduction	Cramer's V	0.300	0.008
Source: Compiled by authors			

Table 7. Internet connection at inter-organizational level- Organizational results.

Internet connection – Performance		Value	Approx. Significance
Internet connection - Market share increase	Cramer's V	0.190	0.508
Internet connection - Increase of profits	Cramer's V	0.225	0.161
Internet connection - Strategic renewal	Cramer's V	0.283	0.006
Internet connection - Efficiency increase	Cramer's V	0.201	0.378
Internet connection - Sales growth	Cramer's V	0.273	0.047
Internet connection - Fulfilment of objectives	Cramer's V	0.309	0.003
Internet connection - Development of new services	Cramer's V	0.175	0.678
Internet connection - Customers loyalty increase	Cramer's V	0.322	0.001
Internet connection - Innovation	Cramer's V	0.198	0.408
Internet connection - Quality improvement	Cramer's V	0.199	0.397
Internet connection - Competitive position improvement	Cramer's V	0.231	0.126
Internet connection - Costs reduction	Cramer's V	0.162	0.805
Source: Compiled by authors			

Table 8. Email – Organizational results.

Email -Performance		Value	Approx. Significance
Email - Market share increase	Cramer's V	0.199	0.398
Email - Increase of profits	Cramer's V	0.215	0.236
Email - Strategic renewal	Cramer's V	0.271	0.013
Email - Efficiency increase	Cramer's V	0.243	0.070
Email - Sales growth	Cramer's V	0.240	0.226
Email - Fulfilment of objectives	Cramer's V	0.284	0.012
Email - Development of new services	Cramer's V	0.244	0.064
Email - Customers loyalty increase	Cramer's V	0.246	0.082
Email - Innovation	Cramer's V	0.224	0.193
Email - Quality improvement	Cramer's V	0.213	0.276
Email - Competitive position improvement	Cramer's V	0.183	0.581
Email - Costs reduction	Cramer's V	0.155	0.859
Source: Compiled by authors			

Table 9. Website- Organisational results.

Website – Performance		Value	Approx. Significance
Website - Market share increase	Cramer's V	0.199	0.404
Website - Increase of profits	Cramer's V	0.181	0.611
Website - Strategic renewal	Cramer's V	0.220	0.194
Website - Efficiency increase	Cramer's V	0.261	0.025
Website - Sales growth	Cramer's V	0.218	0.449
Website - Fulfilment of objectives	Cramer's V	0.266	0.032
Website - Development of new services	Cramer's V	0.219	0.206
Website - Customers loyalty increase	Cramer's V	0.290	0.008
Website - Innovation	Cramer's V	0.290	0.008
Website - Quality improvement	Cramer's V	0.224	0.192
Website - Competitive position improvement	Cramer's V	0.216	0.232
Website - Costs reduction	Cramer's V	0.203	0.352
Source: Compiled by authors			

Table 10. CRM-Organizational results.

CRM-Performance		Value	Approx. significance
CRM - Market share increase	Cramer's V	0.333	0.001
CRM - Increase of profits	Cramer's V	0.289	0.018
CRM - Strategic renewal	Cramer's V	0.311	0.004
CRM - Efficiency increase	Cramer's V	0.269	0.060
CRM - Sales growth	Cramer's V	0.254	0.102
CRM - Fulfilment of objectives	Cramer's V	0.264	0.101
CRM - Development of new services	Cramer's V	0.317	0.002
CRM - Customers loyalty increase	Cramer's V	0.276	0.061
CRM – Innovation	Cramer's V	0.258	0.132
CRM - Quality improvement	Cramer's V	0.303	0.015
CRM - Competitive position improvement	Cramer's V	0.203	0.619
CRM - Costs reduction	Cramer's V	0.298	0.010
Source: Compiled by authors			

- The Central Reservation Systems (CRS), an application of central location, that controls and updates the inventory of rooms, prices in different distribution channels and allows the automatic management of the reserves, produced a large number of organizational results that can be obtained through its use at inter-organizational level. Specifically, most of them, with exception of the development of new services, innovation and competitive position improvement. It indicates that there is a clear orientation towards financial returns (Table 11). According to Figueroa [25], the CRS offer clear advantages for the management of a hotel which is also supported by our study.

- Finally, more innovative technologies used with stakeholders, especially customers, namely, the Web 2.0 and the blogs, also have a significant association with most of the analysed results unless with profits increase, strategic renewal, competitive position improvement and costs reduction,. i.e., results closely linked with the financial aspects of the company (Table 12).

5. CONCLUSIONS AND DISCUSSION

According to results of descriptive statistics, we find that Internet is the most valued technology by respondents from the analysed hotels because of its utility at both intra-organizational learning - learning among employees, as well as inter-organizational level-learning with the stakeholders. These results are in line with Sooraksa [65] who indicates that organizational learning among employees in the hotel industry has involved extensive use of Internet. Furthermore, Internet is important in the commercial area referring to a good learning from customers, suppliers, etc. as testified by Redondo [66].

Table 11.CRS-Organizational results

CRS-Performance		Value	Approx. Significance
CRS - Market share increase	Cramer's V	0.310	0.004
CRS - Increase of profits	Cramer's V	0.289	0.018
CRS - Strategic renewal	Cramer's V	0.348	0.000
CRS - Efficiency increase	Cramer's V	0.309	0.004
CRS - Sales growth	Cramer's V	0.273	0.033
CRS - Fulfilment of objectives	Cramer's V	0.281	0.048
CRS - Development of new services	Cramer's V	0.265	0.074
CRS - Customers loyalty increase	Cramer's V	0.368	0.000
CRS - Innovation	Cramer's V	0.239	0.256
CRS - Quality improvement	Cramer's V	0.334	0.002
CRS - Competitive position improvement	Cramer's V	0.248	0.165
CRS - Costs reduction	Cramer's V	0.395	0.000
Source: Compiled by authors			

We can find studies focused on how organizational learning can affect organizational effectiveness in the tourism sector [7-8] and specifically, in the hotel sector it is proved empirically that organizational learning influences organizational performance [67]. Then, some researchers analyse organizational learning as a cause that explains and solves the problems related to new technologies use in organizations [3]. We analysed which technologies associated with learning generate specific organizational results. We found that the most valued result is customer loyalty increase.

Specifically, organizational learning-oriented technologies which have had a greater effect on organizational outcomes on intraorganizational level are Internet and Intranet. Internet use among hotel employees assumes, according to our study, an improvement of all organizational results analysed unless operational efficiency increase and costs reduction. This means that Internet is a very powerful tool and a key for the development of appropriate management of the tourist organization through organizational learning, as indicated by Martin-Rojas et al. [5].

Table 12. Web 2.0/Blogs-Organizational results.

Web 2.0/Blogs-Performance		**Value**	**Approx. Significance**
Web 2.0/Blogs - Market share increase	Cramer's V	0.282	0.028
Web 2.0/Blogs - Increase of profits	Cramer's V	0.270	0.056
Web 2.0/Blogs - Strategic renewal	Cramer's V	0.237	0.254
Web 2.0/Blogs - Efficiency increase	Cramer's V	0.274	0.045
Web 2.0/Blogs - Sales growth	Cramer's V	0.273	0.032
Web 2.0/Blogs - Fulfilment of objectives	Cramer's V	0.349	0.001
Web 2.0/Blogs - Development of new services	Cramer's V	0.310	0.004
Web 2.0/Blogs - Customers loyalty increase	Cramer's V	0.394	0.000
Web 2.0/Blogs - Innovation	Cramer's V	0.371	0.000
Web 2.0/Blogs - Quality improvement	Cramer's V	0.282	0.047
Web 2.0/Blogs - Competitive position improvement	Cramer's V	0.231	0.312
Web 2.0/Blogs - Costs reduction	Cramer's V	0.257	0.108
Source: Compiled by authors			

Although the use of Intranet supports the use and access to knowledge flowing in the organization [68-71], this study has shown that even though it may influence some organizational results, it has no such a positive impact as Internet.

In order to obtain positive results, the most valued technologyoriented to organizational learning at inter-organizational level is the CRS, followed by Web 2.0, blogs and CRM and to a lesser extent Internet, website and email which can be highlighted as they are technologies closely linked to customers and suppliers although two of them, Internet and website, are significantly associated with customers loyalty. The CRS allows mainly to control, promote and sell products globally as well as to facilitate the yield management, as advocates Buhalis [26]. This can be related mainly to good financial returns, as indicated in this study, through results such as market share increase, profits increase, sales growth, etc.

Web 2.0 and blogs have also been highlighted in this study as a technology that can generate efficient results if used as a learning tool at inter-organizational level. According to Lim et al. [72] introducing Web 2.0 technologies allow to interact with clients to share comments and provide feedback.

On the one hand, from an academic perspective, this study enhances the understanding of the importance of ICTs as a tool that can improve organizational results from a learning point of view. As mentioned in the literature, previous research has been more focused on the introduction and development of ICTs and their costs and benefits [16-19,21], but not from a specific point of view as it is the learning. On the other hand, this study re-emphasize that Internet can improve performance and competitiveness, as indicated also by Namasivayam et al. [73], Porter [74] and Salter et al. [75]; and the CRS could be seen as a new competitive source, as indicated by Mihalič and Buhalis [14].

From the business perspective, we conclude that (1) the use of Internet is a key learning tool that can generate positive results in hotels and, above all, its use at inter-organizational level, it is to say, between employees. It means that employees have to make more use of it like Intranet (2) that the hotels should promote the use of the CRS as interaction tool with stakeholders as it can generate effective results at inter-organizational level similarly as the blogs and Web 2.0; (3) that the customer loyalty increase is one of the most appreciated results from the use of ICT as a learning tool. Thus, hotels should give it a boost as customer loyalty can lead to better financial results.

We have to point out that there are a number of limitations in this work. First of all, the measurement of variables is based on the subjective assessments of respondents. The collected information comes from the perceptions of a single informant - senior and middle managers of the hotel establishments, which includes a certain degree of subjectivity in their opinions. Even though it is a common approach in such studies, the possibility of including more objective measures to improve the validity of the obtained results could be considered. However, it is not easy to obtain such information in the hotel industry which makes difficult gathering more objective measurements.

Finally, we could point out the following future research lines: (1) it would be interesting to consider a qualitative research methodology, introducing case studies, in order to be able to compare between our results and business practices; (2) owing to the fact that our study includes hotels of 3, 4 and 5 stars, a comparison between the level of learning achieved by different categories could be made. Furthermore, a cluster analysis could be carried out according to the size of the hotel, age, number of employees and belonging or not to a chain; (3) finally, this study could be carried out in other Spanish regions or regions in other countries in order to determine whether there are significant statistical differences in comparison with our results of the ICTsrelated organizational results in terms of learning.

REFERENCES

1. Law R, Leung R, Au N, Lee HA (2013) Progress and development of information technology in the hospitality industry: evidence from Cornell hospitality quarterly. Cornell Hospitality Quarterly 51: 10-24.

2. Law R, Buhalis D, Cobanoglu C (2014) Progress on information and communication technologies in hospitality and tourism. International Journal of Contemporary Hospitality Management 26: 727-750.

3. Robey D, Boudreau MC, Rose GM (2000) Information technology and organizational learning. A review and assessment of research. Accountingmanagement and Information Technologies 10: 125-155.

4. Mihi-Ramirez A, Garcia Morales VJ, Martin Rojas R (2011) Knowledge creation, organizational learning and their effects on organizational performance. Inzinerine Ekonomika-Engineering Economics 22: 309-318.

5. Martín-Rojas R, García-Morales, VJ, Mihi-Ramírez A (2014) Knowledge-based organization in tourism industry. InzinerineEkonomika-EngineeringEconomics 25: 82-93.

6. DiBella A, Nevis EC, Gould JM (1996) Understanding organizational learning capability. Journal of Management Studies 33: 361-379.

7. Petrash G (1996) Dow´s journey to a knowledge value management culture. European Management Journal 14: 365-373.

8. Gupta AK, GovindarajanV (2000) Knowledge flows within multinational corporations. Strategic Management Journal 21: 473-496.

9. Olivera F (2000) Memory systems in organizations: an empirical investigation of mechanisms for knowledge collection. Storage and access. The Journal of Management Studies 37: 811-832.

10. Ruíz-Molina ME, Gil-Saura I, Moliner-Velázquez B (2011) Does technology make a difference? Evidence from Spanish Hotels. Service Business. An International Journal 5: 1-12.

11. Law R, Jogaratnam G (2005) A study of hotel information technology applications. International Journal of Contemporary Hospitality Management 17: 170-180.

12. Piccoli G, Lui T-W (2014) The competitive impact of information technology: can commodity IT contribute to competitive performance? European Journal of Information Systems 23: 616-628.

13. Terrett-Rojas R, García Morales VJ, Mihi Ramirez A (2011) How can we increase Spanish technology firms' performance? Journal of knowledge Management 15: 759-778.

14. Mihalič T, Buhalis D (2013) ICT as a new competitive advantage factor-Case of small transitional hotel sector. Economic Business Review 15: 33-56.

15. Sirirak S, Islam N, Khang DB (2011) Does ICT adoption enhance hotel performance? Journal of Hospitality and Tourism Technology 2: 34-49.

16. Hudson BT (1994) Innovation through acquisition. The Cornell Hotel and Restaurant Administration Quarterly 35: 82-87.

17. Jones P (1996) Management hospitality innovation. The Cornell Hotel and Restaurant Administration Quartely 37: 86-97.

18. Van Hoof HB, Verbeteen MJ, CombrickTE (1996) Information technology revisited-international lodging-industry technology needs and perceptions: A comparative study. The Cornell Hotel and Restaurant Administration Quarterly 37: 86-91.

19. Feinstein AH, Stefanelli JM (1999) Technology application in foodservice pursaching: A primer for foodservices marketers. Journal of Restaurant and foodservice Marketing 32: 3-22.

20. Leung R, Law R (2007) Information technology publications in leading tourism journals: a study of 1985 to 2004. Information Technology and Tourism 9: 133-144.

21. Yang J-T (2008) Individual attitudes and organizational knowledge sharing. Tourism Management 29: 345-353.

22. Law R, Leung R, Buhalis D (2009) Information technology applications in hospitality and tourism: a review of publications from 2005 to 2007. Journal of Travel and Tourism Marketing 26: 599-623.

23. Ip C, Leung R, Law R (2011) Progress and development of information and communication technologies in hospitality. International Journal of Contemporary Hospitality Management 23: 533-551

24. XiaoqiuMaJ, Buhalis D, Song H (2003) ICTs and Internet adoption in China´s Tourism Industry. International Journal of Information Management 23: 451-467.

25. Figueroa C, Talón P (2007) La gestión de la información en el sector hotelero madrileño: La inversión en TICs. La importancia de la conectividad y las características del hotel. XXI Congreso Anual AEDEM. Universidad Rey Juan Carlos. Madrid.

26. Buhalis D, Main H (1998) Information technology in peripheral small and medium hospitality enterprises: Strategic analysis and critical factors. International Journal of Contemporary Hospitality Management 10: 198-202.

27. Xiang Z, Gretzel U (2010) Role of social media online travel information search. Tourism Management 31: 179-188

28. Ryssel R, Ritter T, Gemunden HG (2004) The impact of information technology deployment on trust. Commitment and value creation in business relationships. Journal of Business and Industrial Marketing 19: 197-207.

29. Fiol C, Lyles M (1985) Organizational learning. Academy of Management Review 10: 803-813.

30. Spender JC (2008) Organizational Learning and Knowledge Management: Whence and Whither? Management Learning 39: 159-176.

31. Moreno MJ, Vargas A (2003) El modelo THALEC para la gestión interna del conocimiento: Una definición de los elementos facilitadores del aprendizaje organziativo. XII Congreso de la Asociación Española de Contabilidad y Administración de Empresas (AECA). Cádiz.

32. Terrett A (1998) Knowledge management and the law firm. Journal of Knowledge Management 2: 67-76.

33. Mazón C, Pereira P (1999) Las empresas industriales y las tecnologías de Internet. Economía Industrial 329: 99-108.

34. Kahle E (2002) Implications of new economy traits for the tourism industry. Journal of Quality Assurance in Hospitality and Tourism 3: 5-23.

35. Hjalager A (2002) Repairing innovation defectiveness in tourism. Tourism Management 23: 465-474.

36. Rodrígues J, Burguet G (2003) El turismo. La ciencia y la innovación tecnológica. Mundoturístico 2: 2-4.

37. Buhalis D, Law R (2008) Progress in information technology and tourism management: 20 years on and 10 years after internet. The state of e-tourism research. Tourism Management 24: 609-623.

38. Martínez León I (2002) El aprendizaje en las organizaciones. Aplicación al sector agroalimentario. Tesis Doctoral. Universidad Politécnica de Cartagena.

39. Piccoli G, O´Connor P, Capiccioli C, ÁlvarezR (2003) Customer relationship management: A driver for change in the structure of the US lodging industry. Cornell Hotel and Restaurant Administration Quartely 44: 61-73.

40. Ham S, Kim W, Jeong S (2005) Effect of information technology on performance in upscale hotels. International Journal of Hospitality Management 24: 281-294.

41. Siguaw A, Enz A, Namasivayam K (2000) Adoption of information technology in US hotels strategically driven objetives. Journal of Travel Research 39: 192-201.

42. Sahadev S, Islam N (2005) Why hotels adopt ICTs: A study on the ICT adoption propensity of hotels in Thailand. International Journal of Contemporary Hospitality Management 17: 391-401.

43. Brown J, Dev C (1999) Looking beyond RevPar: Productivity consequences of hotel strategies. Cornell Hotel & Restaurant Administration Quarterly 40: 23-33.

44. SigalaDaM (2003) The information and communication technologies productivity impact on the UK hotel sector. International Journal of Operations & Productions Management 23: 1224-1245.

45. Witt AC, Witt F (1989) Why productivity in the hotel sector is low? International Journal of Contemporary Hospitality Management 1: 28-33.

46. David J, Grabski. S, Kasavana M (1996) The productivity paradox of hotel-industry technology. The Cornell Hotel & Restaurant Administration Quarterly 37: 64-70.

47. Olsen MD and Connolly. D.J. (2000) Experience-based travel: How technology is changing the hospitality industry. Cornell hotel and Restaurant AdministrationQuarterly 40: 30-40

48. Tae GK, Jae HL, Rob L (2008) An empirical examination of the acceptance behavior of hotel front office systems: An extended technology acceptance model. Tourism Management 29: 500-513.

49. Karadag E, Cobanoglu C, Dickinson C (2009) The characteristics of IT investment decisions and method used in the US lodging industry. International Journal of Contemporary Hospitality Management 21: 52-68.

50. INE (2014) Encuesta de ocupación hotelera. Principales resultados 2014 provisional.

51. Comisión Europea (2003) CE. Diario Oficial de la UE.

52. Wan H (2000) Opportunities to enhance a commercial web site. Information & Management 38: 15-21.

53. Garbin D (2008) The effects of Intranet. Internet and Extranet on performance in hospitality industry: A case of split region. An enterprise Odyssey. International Conference Proceedings: 860-873. Zagreb.

54. Stevens RG, McElhill J (2000) A qualitative study and model of the use of e-mail in the organizations. Internet research: Electronic Networking Applications and Policy 10: 271-283.

55. GarauVadell JB, Orfila-Sintes F (2008) Internet innovation for external relations in the Balearic hotel industry. The Journal of Business & Industrial Marketing 23: 70-80.

56. Frías DM, Rodriguez MA, Castañeda JA (2008) Internet vs. travel agencies on pre-visit destination image formation: An information processing view. Tourism Management 29: 163-179.

57. Bonn MA, Furr HL, Susskind AM (1988) Using the Internet as a pleasure travel planning tool: An examination of the sociodemographic and behavioural characteristics among Internet users and non-users. Journal of Hospitality & Tourism Research 22: 303-317.

58. Luo S, Lorenzi F, Saldana R, Licthnow D (2004) A tourism recommender system based on collaboration and next analysis. Information Technology and Tourism Marketing 17: 15-25.

59. O´Connor P (2008) E-mail marketing by international hotel chains. Cornell Hospitality Quartely 49: 42-52.

60. Bai B, Law R, Wen I (2008) The impact of website quality on customer satisfaction and purchase intentions: Evidence from Chinese online visitors. International Journal of Hospitality Management 27: 391-402.

61. Chu R (2001) What online Hong Kong travellers look for on airline/travel websites? International Journal of Hospitality Management 20: 95-100.

62. Chadha A (2015) Case study of hotel Taj in the context of CRM and customer retention. Kuwait Chapter of Arabian Journal of Business and Management Review 4: 1-9.

63. Reinartz WJ, Krafft M, Hoyer WD (2004) The customer relationship management process: its measurement and impact on performance. Journal of Marketing Research 41: 293-305.

64. Jayachandran S, Sharma S, Kaufman P, Raman. P (2005) The role of relational information processes and technology use in customer relationship management. Journal of Marketing 69: 177-192.

65. Sooraksa N (2012) Using computer-based continuing professional education of training stafftodevelop small- and medium-sized enterprises in Thailand. International Journal of training and Development 16: 235-242.

66. Pablo Redondo R (2004) Las nuevas tecnologías aplicadas al sector turístico. Editorial Universitaria Ramón Areces.

67. Yang J-T (2010) Antecedents and consequences of knowledge sharing international tourist hotels. International Journal of Hospitality Management 29: 42-52.

68. Cohen D (1998) Toward a knowledge context: Report on the first annual U.C. Berkeley forum on knowledge and the firm. California Management Review 40 (3) Spring: 22-39.

69. Ruggles R (1998) The state of the notion: Knowledge management in practice. California Management Review 40: 80-89.

70. Gottschalk P (2000) Strategic knowledge networks: The case of it support for eurojurist law firms in Norway. International Review of Law. Computers&Technology 14: 115-129.

71. Guadamillas Gómez F (2001) La gestión del conocimiento como recurso estratégico en un proceso de mejora continua. Alta Dirección. nº 217 mayo-junio: 199-209.

72. Lim S, Zegarra-Saldaña A, Zegarra-SaldañaP (2011) Do market oriented firms adopt web 2.0 technologies? An empirical study in hospitality firms. International Entrepreneurship and Management Journal 7: 465-477.

73. Namasivayam K, Enz CA, Siguaw JA (2000) How wired are we? Selection and use of new technology in U.S. hotels. The Cornell Hotel and Restaurant Administration Quarterly 41: 40-48.

74. Porter ME (2001) Strategy and the Internet. Harvard Business Review 79: 63-78

75. Slater SF, Narver JC (1995) Market orientation and the learning organization. Journal of Marketing 59: 63-74.

2

LEVERAGING SUSTAINABILITY CERTIFICATION IN CANADA'S URBAN HOTEL MARKETS

Taillon J*, Laurie A and Yhip G

School of Hospitality, Food and Tourism Management,
University of Guelph, Canada

ABSTRACT

The greening of the hotel industry has been identified as important to the industry for a multitude of reasons. Most importantly, being identified as a green property may lead to enhanced profits. Milton Friedman's 1962 property rights argument, the theoretical framework used in this study, stated the sole purpose of an organization such as a hotel is profit. Conveying a property's green qualities is difficult though. Certification has been identified as one important method for conveying a property's green attributes. Yet, the role green certification can play in a hotel's profit is not known. This study seeks to identify the profits gained from earning a specific green certification: Green Key Global. This certification program is North America's most popular with more than 3,000 participating hotels. The data used to make inferences regarding profitability stemming from Green Key certification has been gleamed from recent STR and HVS data. The data focuses on the largest hotel markets in Canada: Toronto, Montreal, and Vancouver. The upper upscale and luxury properties in this city are compared using STR metrics (e.g. occupancy, supply, demand, Rev PAR, and ADR).

KEYWORDS

Big data; Certification; Green hotels; Smith travel research; HVS global valuation services; sustainability

1. INTRODUCTION

The greening of the hotel industry is important. First, some studies have shown hotels accepting an environmental stewardship narrative have shown increased profitability; such studies include industry partners Travel Industry Association of America, Orbitz, Travelocity, and Cone Roper [1]. Secondly, customers have shown they consider environmental factors when making hotel decisions [2]. This has also been a questioned discourse; studies have shown consumers state they consider environmental factors but these customers cannot necessarily differentiate green-washing from environmental sensibility and/or they may not be willing to pay more for such environmentally sensitive practices [2]. Finally, employees and managers alike at the property level have stated they are concerned with environmental issues [3].

Green Hotels are environmentally-friendly properties whose managers are eager to institute programs that save water, save energy and reduce solid waste—while saving money—to help PROTECT OUR ONE AND ONLY EARTH! Being green means guests, staff and management are healthier. There's just no doubt that when odors, fumes, soot, droplets and residues of toxic, poisonous chemicals are not in the air, on our food or on anything we touch, we are not absorbing or breathing them. Sooner or later all properties will be sold, and any green property will demand a higher price because it's value is much enhanced by lower utility bills per square foot, its healthier aspects and owner care. The hotel industry forms a large part of the tourism industry and if not managed properly, it has the potential to be detrimental to the social and natural environments within which it functions. Hotels are also resource intensive and in order to reduce their impact, it is imperative to go "green!

Conveying a property's green qualities includes portraying the greenness to all stakeholder groups including customers, ownership, and employees. This has been repeatedly identified as a difficult task. The multitude of options for conveying greenness has lead to "grey" rather than green outcomes. Yet, there are options for conveying green.

For example, Font and Harris identified certification as the best method for conveying a property's green attributes to potential customers.

Certification can provide a seal of quality for visitors [4]. Certification programs can play a role beyond providing a simple ecolabel though.For example, certification programs can impact the community and customer experience if set up with sustainability-oriented foci [5]. Some studies have even identified the implementation of small-scale certification program as beneficial to stakeholders holistically [6].

This study focuses on the application of a specific certification in lieu of certification as a whole. Furthermore, the study focuses on financial attributes of earning certification rather than whether a hotel is green or not. This is because hoteliers are not focused on greening so much as return on investment (ROI) in relation to certification programs. Thus, while a majority of research conducted in hospitality literature is seeking to understand environmental impacts and shortcomings of certification programs this study seeks to understand the financial ROI of earning environmental stewardship certification. No studies have identified the ROI of green certification in a specific program to date. Thus, this study is filling an existing gap in the literature.

Green Key Global is North America's largest hotel certification program. The program has shown fast growth patterns in recent years. This study identifies the correlation between achieving Green Key Global certification and ROI. The data used to identify the correlation, or lack thereof, between properties earning Green Key certification and properties not earning this certification comes from Smith Travel Research (STR), HVS Global Hospitality Services, and Green Key Global. The data from this unique partnership makes this study unique and provides insights that have not previously been published.

2. LITERATURE REVIEW

2.1 Hotel certification

Conveying a property's "green" qualities is difficult. Yet there are opportunities for conveying greening as a CSR component. CSR stands for Corporate Social Responsibility. This includes marketing [7], community outreach, certification [8], and others. This paper will focus on certification. Certification programs have been identified as the primary method of designating a property as "green" and thus conveying environmental stewardship to visitors. Yet, problems persist within hotel certification. Chan and Wong conducted a study that concluded 36% of hotel managers in their study (the study took place in China, Macau, and Hong Kong) had only a minimal or complete lack of understanding of different greening certifications available to them. The authors attempted to establish motivations for implementing, or not participating in conjunction with, specifically the ISO's 14.000 EMS standards. This is a mainstream and generally respected greening program with a lengthy history.

Chan and Wong found that there was little external pressure from upper level management, employees, national and/or local governments, or market conditions to earn certification. The largest issue was that there was no regulating force pushing for compliance with EMS ISO 14.000's or other certifications' standards. Chan and Wong formed a belief that there exists an internal motivation in executive committee managers to create more environmentally sustainable practices in the industry. Yet, the data also shows that the management responsible for implementation, the front-line managers, lack motivation to follow through with the ideas

EMS ISO 14.000 Series is only one certification program. There are more than 400 certification programs available worldwide and approximately 100 available to North American hoteliers. This multitude of options makes choosing the best one for an individual property difficult. Furthermore, each certification program has benefits and drawbacks when an organization chooses to earn the certification. For example, Chan's 2013 study focused on hotel manager attitudes regarding marketing green attributes of hotels to customers. The study found that managers of green-certified hotels believe customers are willing to pay a premium for tangible green services and are less suspicious of a hotel's green claims when the property has earned certification of its green practices. Conversely, managers of un-certified hotels were identified as individuals who believe the benefit of being green-certified is over-stated. Obviously, each group had a bias towards certification that aligned with their property's approach to green certification. Yet, Chan and similar studies such as the price premium of an environmentally friendly product [9,10] the Bottom Line of Green is black, have all identified that hotels do see a positive ROI when it is perceived as green. These studies also concluded that the relationship between ROI and certification needs more investigation. The positive correlation is suspected by hotel management but not necessarily true. One issue consistently credited with hindering the implementation of certification programs in the accommodation portion of the hospitality industry is a perceived lack of financial incentive. The issue is not that there are no financial savings. The issue is that the costs outweigh the benefits. Thus, one possibility to explain why certification programs is failing worldwide is the lack of understanding of how certification can affect the "bottom line" of hoteliers' financial situation [1]. This study answers such calls for investigation by investigating the ROI of North America's most popular hotel certification program.

2.2 Green key global

The Hotel Association of Canada (HAC) created Green Key Global in response to market conditions. Specifically, the greening of the hotel industry was a timely topic in Canada in the 1980s and 1990s (e.g. Earth Summit in Rio de Janeiro). HAC, the lobbying entity of Canada's hotel industry, assisted in forwarding Green Key Global to meet market demand and to ensure green regulation would be uniform and specific within the Canadian hotel industry. The ownership of Green Key Global by an industry partner seen in the marketplace as a regulator is a boon to the program as will be discussed shortly. Basically, the ownership model of Green Key Global gives the program a unique edge when compared to existing hotel certification programs. The market conditions that lead to the inception of Green Key have changed minimally. Hospitality is continuing to move toward sustainability. The hotel industry continues to face pressure to develop measurable goals and to apply performance measurement models to assess sustainability progress in business practices [11]. Green Key is an organization seeking to fill this gap. Green Key assists hotels in developing measurement techniques through sustainability reporting, performance measurement, application of green practices, and creation of an internal sustainability culture according to [11]. Today, Green Key Global can be understood as an environmental certification with more than 3,000 participating hotels. The program exists to evaluate and certify primarily hotels in Canada and the United States based on their environmental initiatives [12]. Within this context the goal of Green Key Global is to create a platform for members to leverage organizational CSR activities to support sustainable initiatives and improve overall fiscal performance and community relations [12]. The Green Key Eco-Rating Program assesses the five operational areas of a property and nine areas of sustainable practices (Table 1). The success of the program in Canada allowed for growth into new markets. For example, in the United States Green Key Global shares a partnership with HAC and LRA Worldwide Inc. in order to certify properties in the United States. Green Key Global is also planning growth in Central and South America. Their successes, sustainable business model, and future growth possibilities make them an apt case to study.

2.3 Problematic qualities of hotel certification programs

Green certification has been identified as problematic for the hotel industry. It is arguable that CSR is often not the reason for hotels to adopt green initiatives [13] writes "a positive inclination among hotel managers towards using economic incentives as a strategy to convince senior-level management about investing in energy-efficiency/ conservation measures or energy audits." An example would be the government providing tax incentives for hoteliers to use sustainable technologies. Another economic incentive would be if hotel operators saw an increase in occupancy because of their eco label. Eco-labels also address the psychological disconnect between people and environmental problems in that they increase consumer awareness and provide the opportunity for immediate decision-making regarding environmental action. Eco-labeling programs provide valuable, more complete information to consumers about the environmental effects of their purchase decisions [14]. Eco labels give guests a choice to spend their money on a hotel that operates in a sustainable manner. Although certification programs are much maligned, certification programs such as Green Key Global have had success in growth in recent years, but the motives and accomplishments of such programs have been questioned. Furthermore, consumers have been shown to not understand differences between programs. For example, most consumers are unable to differentiate green certification brands including specifically Green Key Global [15]. A University of California study defined the inability as a lack of brand awareness amongst consumers regarding eco-rating awarding entities [15]. The term "brand awareness" is at least partially problematic because mission statements of certification programs rarely include branding (indicators of success are more often focused on lowering costs such as electric and water). Green Key was one such program specifically mentioned in the aforementioned study [15]. The multitude of eco-rating awarding entities is the cause of the lack of brand awareness. This has shown that consumers take little notice of eco-ratings when making hotel purchasing decisions. Along with the lack of awareness from the consumer, there exists a lack of understanding of the environmental impact that a hotel has on its surrounding natural environment [15] study survey results identified that most consumers lack awareness of the environmental impacts of hotels. Most respondents (62.3%) in this study answered that they are not sufficiently informed about the actual environmental practices in hotels. Because of what has been coined the lack of brand awareness in awarding entities and the lack of knowledge that consumers have about a hotel's environmental impact, it is arguable that consumers are suspicious of the validity of these entities. Consumers are more likely to trust an eco-rating awarding entity if it is operated by the government or by a not-for-profit organization as alluded to earlier in this paper. Government and non-profit organizations are seen as more trustworthy eco-certifiers, compared to hired private agencies, or a hotel's own ecocertification program [15]. Also, the industry's sustainable practices are often viewed as the way to merely achieve their ultimate marketing goal of maximizing profitability [11]. It is seen in academic studies that if consumers are made more aware of the green-initiatives, it is less likely that the initiatives are seen as green-washing. It is likely that in order to attract consumers to green-certified hotels, educational outreach is required in order to most effectively reduce operating costs while simultaneously increasing attractiveness to potential customers. Furthermore, the hotels should provide detailed information about the specific green practices they have undertaken and should highlight the more unique ones. Such unique practices might perhaps be xeriscape gardens, composting efforts, etc. [11].

Table 1. Operational areas of Green Key certification.

Operational areas	Sustainable practices
Corporate environmental management	Energy conservation
Housekeeping	Water conservation
Food and Beverage Operations	Solid waste management
Conference and Meeting Facilities	Hazardous waste management
Engineering	Indoor air quality
	Community outreach
	Building infrastructure
	Land use
	Environmental management

2.4 Theoretical Framework

Hotel certification programs are much maligned in literature for leading to "grey" outcomes rather than "green" outcomes [4]. This is due to investment strategies in the hotel industry. It has been increasingly identified that "green" is customers' social construction rather than evidence-based; basically, if an organization can convince customers they are green then the customer has a higher level of satisfaction with the brand whether the organization has implemented greening practices or not [1]. Thus, a great deal of recent research in hotel greening has focused on ideal sustainability practices, green outcomes of certification, and marketing green to customers. These foci have led researchers to problems with greening in that they are majority grey outcomes being certified [4]. This study would be remiss without identifying the much-maligned hotel certification industry and stakeholder theory. Yet, this paper approaches certification within a different framework. Specifically, we accept Milton Friedman's 1962 property rights argument as the framework for this paper. It is possible that this argument best explains the current status of greening in the hotel industry. Friedman, as interpreted [6], evidenced decisionmakers in an organization who fail to maximize shareholder wealth are violating the shareholder agreement. Thus, organizations such as hotels exist solely to make a profit. This means that greening decisions made by hotels should not exist on a foundation of corporate social responsibility or in the best interests of society as a whole (e.g. stakeholder theory), but rather should be profit-based. Friedman's argument has faced scrutiny, particularly from those claiming hotels have are obligated to be upstanding corporate citizens. Some academics have even attempted to discount the application of Friedman's argument in the context of the hotel industry. For example, Ferrero, Hoffman, and McNulty (2012, p. 1-2): Milton Friedman famously stated that the only social responsibility of business is to increase its profits, a position now known as the shareholder model of business. Subsequently, the stakeholder model, associated with Edward Freeman, has been widely seen as a heuristically stronger theory of the responsibilities of the firm to the society in which it is situated. Friedman's position, nevertheless, has retained currency among many business thinkers. In this paper we argue that Friedman's economic writings assume an economy in which businesses operate under the protections of limited liability, which allows corporations to privatize their gains while externalizing their losses. By accepting limited liability, Friedman must also accept a view of business as embedded in social interdependency, which serves as the logical and moral foundation for corporate social responsibility (CSR). To restore consistency to his economic principles, Friedman must refuse limited liability or modify his doctrine on CSR and the related stakeholder model of business. CSR in the context of this study is not important though. We are investigating profit. According to Friedman organizations such as hotels exist to make a profit. We accept this as the goal of a hotel for the purposes of this study. It is possible that consumers are willing to pay a premium to organizations using what they perceive to be sustainable practices in the accommodation sector. In fact, according to multiple industry studies, as compiled [1], this has proven to be the case. Saunders study shows that Friedman's argument does apply:

1) A travel Industry Association of America study found 54% of Americans are more likely to patronize hotels that practice environmental responsibility than those that do not.

2) An Orbitz study that figured 67% of Americans place importance on the eco-friendliness of an organization when choosing a hotel.

3) A Travelocity study stated 80% of Americans are willing to spend more on an eco-friendly destination and

4) A Cone Roper Study Saunders cited found that 95% of consumers in the accommodation sector of the hospitality industry have a more positive image of a company that supports causes they care about. Based on the aforementioned data it is possible that Milton Friedman's property rights argument applies to hotel certification. We accept this as the framework for our study while understanding we are testing solely the profits garnered from earning certification in the single most accepted green certification program in North America.

3. METHODOLOGY

We identified the largest Canadian hotel markets as the sample population for this study (Vancouver, Toronto, and Montreal). We created comp sets that include all upper upscale and luxury market segments as STR defines them for the city. Further parameters ensured we were only relying on hotels involved in direct competition. This was based on discussions with HVS, STR, and personal knowledge of each city we were interested in better understanding and brand included in the comp sets. We created two comp sets for each city's hotels. One comp set for hotels that have earned Green Key certification and one comp set for hotels that have not earned Green Key certification. No upper upscale or luxury hotel was left out of a comp set unless HVS, STR, and the researchers agreed it could not be considered a direct competitor and/or the property does not subscribe to STR data analytics sharing. Using the aforementioned approach we created two comp sets for each city for a total of six comp sets. The comp sets are as depicted in Table 2 (Green Key certified hotels) and Table 3 (hotels without Green Key certification): We compared the five generally accepted primary indicators of operational financial performance between each comp set (e.g. ADR, Rev PAR, Occupancy, Supply, and Demand). We also compared HVS P&L data for the hotels to ensure we best understood the hotel market and to ensure the STR data was correct. This twopronged approach allowed for multiple points of comparison between each hotel grouping. We viewed P&L statements using HVS provided for as many hotels as possible. This was for three reasons. First, to ensure all hotels were reporting correctly. Secondly, this was used to understand how USALI can allow for greater insights to hotel income statements. Finally, the HVS P&L statements were used to identify correlations between certification and savings. Specifically, we sought to understand whether certification lead to changes in operations that should be included when viewing savings. For example, most hoteliers have changed shower heads, light bulbs, and commodes in order to save money. Perhaps there are other changes brought about and/ or instigated by certification in Green Key that lead to cost savings. These cost savings are claimed to occur [12] but no evidence currently exists that such savings are realized at the property level. The savings [12] claims will occur exist in Corporate Environmental Management, Conference and Meeting Spaces, F&B Outlets, Engineering and Housekeeping. Finally, HVS' P&L statements were used to measure the quantitative and qualitative cost of the five operational areas as defined by Green Key: corporate environmental management, housekeeping, food and beverage operations, conference and meeting facilities, and engineering. HVS already collects this data, but the information has not previously been published or viewed in a way whereby the effects of a specific green certification program investigated. The HVS portion of this study is not approved by HVS for passing along yet due to privacy rules. It should be noted that this study is investigative. It is not intended as the culmination of a major research project. Thus, some of the glaring deficiencies in data will need to be fixed in future studies. There are many limitations to this study though. First, the data used in this study is from a small sample size. Secondly, the sample size is small primarily because we sought to make as direct comparisons as possible within the study. Yet, although the samples are similar, there remain marked differences in each market. Finally, we looked at upper upscale and luxury market segments and only in Canada. Each of these could be an outlier in a greater sample size. A larger sample size is not currently available. STR and HVS have significant restrictions on the data that can be used. Yet, this study is the initial foray into whether certain comp sets see a change in Rev PAR, ADR, Occupancy, Supply, or Demand based on earning green certification.

Table 2. Green Key Certified Hotels.

Property	Rooms	STR Classification
Vancouver		
Westin Bayshore Vancouver	511	Upper Upscale Class
Four Seasons Vancouver	372	Luxury Class
Hyatt Regency Vancouver	644	Upper Upscale Class
Marriott Vancouver Pinnacle Downtown	438	Upper Upscale Class
Pan Pacific Hotel Vancouver	503	Upper Upscale Class
Renaissance Vancouver Harbourside	442	Upper Upscale Class
Sheraton Vancouver Wall Centre	733	Upper Upscale Class
Sutton Place Hotel	561	Upper Upscale Class
Toronto		
InterContinental Toronto Centre	586	Luxury Class
Sheraton Centre Toronto Hotel	1377	Upper Upscale Class
The Westin Harbour Castle	977	Upper Upscale Class
Renaissance Toronto Hotel Downtown	348	Upper Upscale Class
Marriott Toronto Eaton Centre	461	Upper Upscale Class
Montreal		
Delta Montreal	456	Upper Upscale Class
Hyatt Regency Montreal	395	Upper Upscale Class
Hilton Montreal Bonaventure	395	Upper Upscale Class
Marriott Chateau Champlain	611	Upper Upscale Class
Le Centre Sheraton Montreal	825	Upper Upscale Class

Table 3. Non Green Key Certified Hotels.

Property	Rooms	STR Classification
Vancouver		
Preferred Metropolitan Hotel	197	Upper Upscale Class
L'Hermitage Hotel Vancouver	60	Luxury Class
Loden Vancouver	77	Luxury Class
Opus Vancouver	96	Luxury Class
Rosewood Hotel Georgia Vancouver	156	Luxury Class
Wedgewood Hotel Vancouver	83	Luxury Class
Toronto		
King Edward	301	Upper Upscale Class
Ritz Carlton	263	Luxury Class
Four Seasons	259	Luxury Class
Shangri-La	202	Luxury Class
Trump	241	Luxury Class
Montreal		
Omni Mount Royal Montreal	299	Upper Upscale Class
Courtyard by Marriott Montreal	212	Upscale Class
Hotel Le Crystal	131	Luxury Class
Lowes Hotel Vogue	142	Luxury Class
Le Place D'Armes Hotel and Suites	133	Luxury Class

4. RESULTS

The data founded in this study does not show that consumers pay more for a Green Key certified product. The results of this can be seen in Table 4 for the Vancouver comp set since 2009. The data depicted in the table is problematic though due to the large percent difference in supply, the brands (e.g. Green Key hotels are branded and there are independents and boutique hotels in the non-Green Key segment), and trends in year-over-year alterations to the market. It should also be noted that Green Key hotels achieve a higher percent occupancy even with the large supply. Tables 5 depicts a larger segment than just one city it includes Toronto and Montreal as well as Vancouver for the 2013 year. It should be noted the trends in Montreal and Toronto are identical to what is noted in Vancouver's hotel market. Occupancy is higher at Green Key properties by just over 20% in the Toronto market and 5% in the Montreal market. The ADR is $121.32 and Rev PAR is $28.69 better in non-Green Key properties in Toronto while this same data for Montreal is $155.19 to $186.35 for ADR and $107.02 compared to $121.35 for Rev PAR, both in favor of the non-Green Key affiliated properties. Like Vancouver, there is a marked difference of supply in the Toronto market between Green Key and non-Green Key upper upscale and luxury properties.

Table 4. Vancouver Comp Set STR Analysis.

Year	Green Key	Non-GK	Difference
Occupancy (%)			
2009	67.70	64.90	2.80
2010	72.90	70.90	2.00
2011	71.90	66.80	5.10
2012	72.00	69.90	2.10
2013	71.20	70.60	0.60
ADR (CAN $)			
2009	166.91	186.58	-19.67
2010	181.42	209.06	-27.64
2011	175.00	198.24	-23.24
2012	170.06	202.79	-32.73
2013	172.19	210.52	-38.33
RevPAR (CAN $)			
2009	112.95	121.03	-8.08
2010	132.30	148.31	-16.01
2011	125.74	132.39	-6.65
2012	122.37	141.80	-19.43
2013	122.66	148.57	-25.91
Supply			
2009	1,533,365	187,245	1,346,120
2010	1,534,589	187,245	1,347,344
2011	1,534,825	215,765	1,319,060
2012	1,534,733	244,185	1,290,548
2013	1,534,460	244,185	1,290,275

Demand			
2009	1,037,630	121,461	916,169
2010	1,119,057	132,838	986,219
2011	1,102,814	144,091	958,723
2012	1,104,344	170,738	933,606
2013	1,093,045	172,331	920,714
Revenue (CAN $)			
2009	173,188,987	22,662,610	150,526,377
2010	203,024,386	27,771,060	175,253,326
2011	192,990,279	28,565,149	164,425,130
2012	187,805,686	34,624,218	153,181,468
2013	188,212,750	36,278,488	151,934,262

Finally, it should be noted that the data trends are identical in all three markets. This is a small sample size, but larger sample sizes in the Canadian urban market are not possible. These are the three largest markets and are thus naturally representative. It is clear that occupancy is higher at Green Key hotels. It is also clear that Rev PAR and ADR are not higher at Green Key hotels. This is perhaps aligned with supply figures.

5. RECOMMENDATIONS AND CONCLUSION

Green certification in the hotel industry is important. The certification is young and being molded currently though. There are more than 100 options in North America for certification. This has confused consumers. Consumers are likely not making purchasing decisions based on green certification currently, at least not Green Key certification. Yet, the greening of the industry has been identified as important. Generation Y turnover figures are lower when they believe a hotel company is green, brands have a higher marketplace awareness, and most importantly, electricity, water, and other metrics are improved upon when meeting certification standards. The data in this study is consistent: occupancy is higher at Green Key properties but Rev PAR and ADR are lower. Hoteliers should focus on being sustainable and earning certification for reasons beyond ADR and Rev PAR growth.

Table 5. City markets, 2013.

City	Green Key	Non-GK	Difference
Occupancy (%)			
Toronto	78.30	55.70	22.60
Vancouver	71.20	70.60	0.60
Montreal	69.00	65.10	3.90
AVERAGE	**72.83**	**63.80**	**9.03**
ADR (CAN $)			
Toronto	171.87	293.19	-121.32
Vancouver	172.19	210.52	-38.33
Montreal	155.19	186.36	-31.17
AVERAGE	**166.42**	**230.02**	**-63.61**
RevPAR (CAN $)			
Toronto	134.51	163.20	-28.69
Vancouver	122.66	148.57	-25.91
Montreal	107.02	121.38	-14.36
AVERAGE	**121.40**	**144.38**	**-22.99**
Supply			
Toronto	1368385	463020	905365
Vancouver	1534460	244185	1290275
Montreal	1,055,580	283,189	772,391
AVERAGE	**1319475**	**330131**	**989344**
Demand			
Toronto	1070935	257728	813207
Vancouver	1093045	172331	920714
Montreal	727,926	184,445	543,481
AVERAGE	**963969**	**204835**	**759134**
Revenue (CAN $)			
Toronto	184,056,454	75,563,090	108,493,364
Vancouver	188,212,750	36,278,488	151,934,262
Montreal	112,969,629	34,372,343	78,597,286
AVERAGE	**161,746,278**	**48,737,974**	**113,008,304**

REFERENCES

1. Saunders T (2009) The Bottom Line of Green Is Black. Paper presented on behalf of the Saunders Hotel Group, Boston, MA.

2. Baker MA, Davis E, Weaver P (2014) Eco-friendly attitudes, barriers to participation and differences in behavior at green hotels. Cornell Hospitality Quarterly 55: 89-99.

3. Chan E (2013) Managing green marketing: Hong Kong hotel Managers. Perspective. International J. of Hosptly Mgmt 34: 442-461.

4. Buckley R (2009) Ecotourism: Principles and Practices. Oxford shire, UK: CABI Tourism Texts.

5. Whitehead M (2007) Spaces of Sustainability: Geographical Perspectives on the Sustainable Society. England: Routledge Press, London.

6. Phillips M, Pegas F (2009) Review of guide training and certification programs. Corpus Christi CVB Working Paper.

7. Lee JS, Hsu LT, Han H, Kim Y (2010) Understanding how consumers view green hotels: How a hotel's green image can influence behavioral intentions. J. of Sustainable Tourism 18: 901-914.

8. Dodds R, Joppe M (2005) CSR in the tourism industry? The status of and potential for certification, codes of conduct, and guidelines. Foreign Investment Advisory Service Investment Climate Department.

9. Kapelianis D, Strachan S (1996) The price premium of an environmentally friendly product. South African J. of Busi Mgmt 27: 89-96.

10. Laroche M, Bergeron J, Forleo GB (2001) Targeting consumers who are willing to pay more for environmentally friendly products. J. of Consu Marktg 18: 503-520.

11. Jayawardena C, Pollard T, Chort V, Choi C, Wanjohi K (2013) Trends and sustainability in the Canadian tourism and hospitality industry. Worldwide Hospitality and Tourism Themes 5: 132-150.

12. Green Key Global (2014) Global Leader in Green Hotels and Meeting Rooms.

13. Halbe A (2013) Green energy initiatives in the hotel industry: Factors influencing adoption decisions.

14. Anderson L, Mastrangelo C, Chase L, Kestenbaum D, Kolodinsky J (2013) Eco-labeling motor coach operators in the North American travel tour industry: Analyzing the role of tour operators. J. of Sustainable Tourism 21: 750-764.

15. Jhawar A, Kohli G, Li J, Modiri N, Mota V, et.al. (2012). Eco-certification programs for hotels in California: Determining consumer preferences for green hotels. Walt Disney Imagineering White Paper, Los Angeles, California.

.3

FLEXIBLE, MULTI-PHASED HOTEL DEVELOPMENT

Roman Milwicza * Sebastian Dubasb

a,b Institute of Structural Engineering, Faculty of Civil and Environmental Engineering, Poznan University of Technology, 5, Piotrowo Str., PL-60-965 Poznan, Poland

ABSTRACT

There is a need for diverse development strategy considering risks of investing in the hospitality market adequate for time varying market requirements. Implementation of long-term investments based on assumptions and forecasts causes a high risk of loss. The target is to find solutions preventing risks and seizing opportunities. In a case study hotel expansion is planned therefore different scenarios were analysed. Historical data enable prediction of market changes; as a result spectrum of demand for services was created. Standard and flexible approach to selected scenarios was analysed with emphasis on development of investment process, minimize risk and optimize profit.

KEYWORDS

flexible design; hotel development; flexible adjustment

1. INTRODUCTION

Nowadays pace of changes accelerates very quickly. Uncertainties become more and more visible even for regular person. Such a situation forces designers and managers to change way of thinking, designing and management. Forecast based on historical data can lead to wrong future decisions. Entrepreneurs are more conscious about importance of designing phase in future gains. Thus, there is a need to change paradigm of designing from stiff-traditional to modern-future proof based on flexibility. Hotel industry, especially in regions located far from urbanized zones is greatly under influence of uncertainty. The article investigates a small-scale hotel case. Implementation of new investment development designing based on demand and potential of the services is shown. Random sampling method was applied in order to simulate future possibilities of hostility demand.

2. FLEXIBILITY

2.1. Future

Forecast Future is unknown. There is no method to forecast true conditions in the future. The longer forecast is the less precise results are obtained. Preparing numerous simulations based on randomly reaching into past historical data gives various trends. Fig. 1 shows that for studied case difference in predictions based on 4 and 6 year historical data reaches almost 36%. What is more, when data gives no regular pattern and coefficient of determination gives low values, trend line predictions may be irrelevant. Such an activity leads to spread of solutions and in traditional approach most probable scenario needs to be selected. Example of divergence forecasting from reality is presented by the oil prices in Fig. 2. As de Neufville [8] says forecast is "always wrong". Historical data is easily accessible, historical forecasts after the fact are very seldom published due to their inaccuracy, hence investigation of past forecasts is still insufficiently developed.

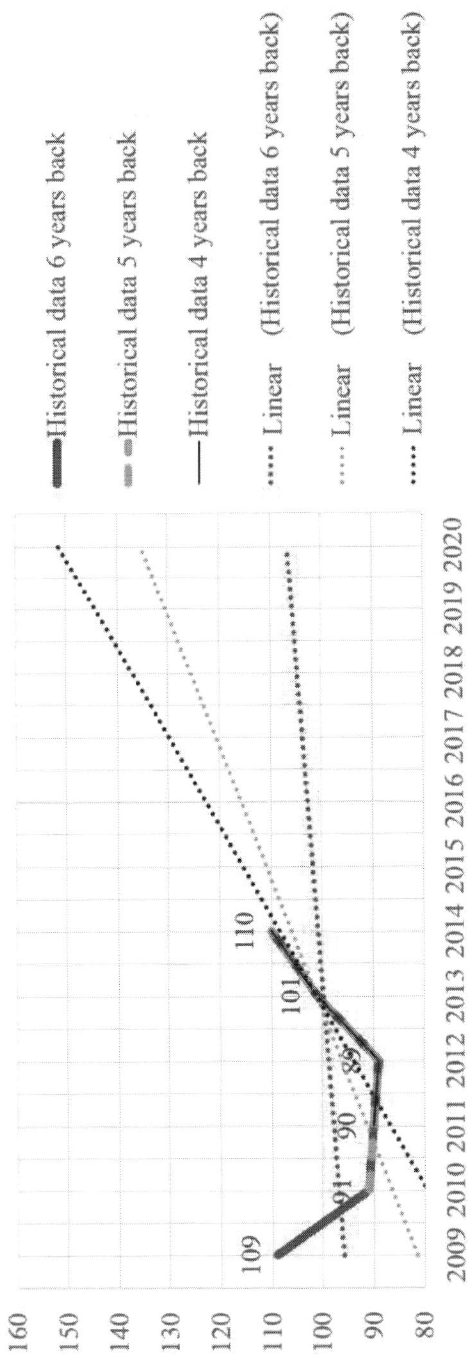

Figure 1. Number of days with over 90% of occupancy (peaks).

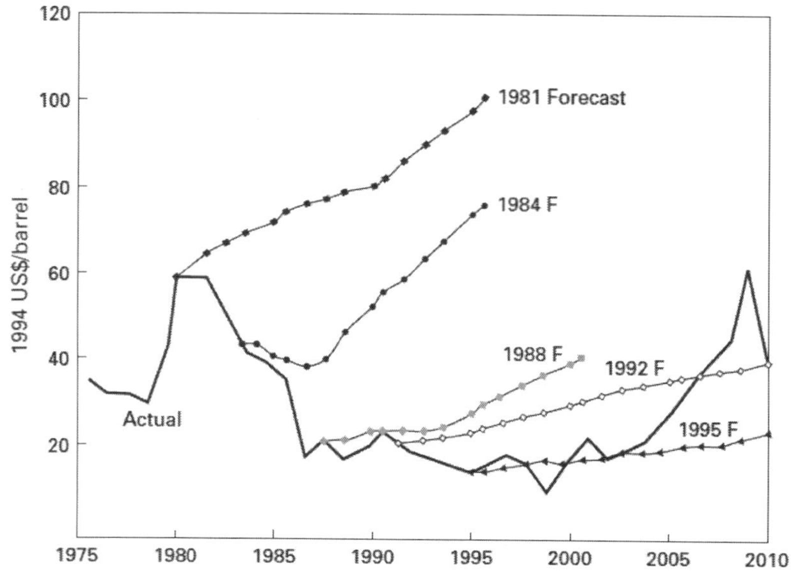

Figure 2. Crude oil price and price forecast [20].

2.2. Flaw of averages

For many years most tools were based on the average value. In recent history changes are so significant that standard deviation is high. Very often observation of one parameter can lead to wrong evaluation of situation. In most cases extreme values generate the need of changes or intervention for example in hotel industry- the need for more rooms is generated by peak season. Due to Jensen's inequality for convex function (generally met in developing enterprise) function of average value is smaller or equal to average value of the function (formula 1). It can mean that it is better to prepare approximation with polynomial or polyline rather than linear function to achieve more realistic results. However finding appropriate equation describing actual future events is always a question.

$$f(E[x]) \leq E[f(x)] \quad (1)$$

2.3. Definition of flexibility

Occurrence of flexible solution can be found in area of life, thus it leads to multiplicity of definitions. Let us concentrate on management branch. Many authors describe cases about strategic management in organizations [3, 14]. Their main goal is focused on client expectation, but in case of civil engineering the importance is shifted towards environmental aspects. On the other hand in hospitality industry the validity is balanced between economy customer and environment. As Osipova and Eriksson [2] say: formal tools deal with identified risks; flexibility copes with unforeseen events. Browne and Sethi defined 11 types of flexibility [4, 9], the most important can be named: expansion flexibility - the ability to build out the capacity of a system; market flexibility - the ability of the system to adapt to market demands; product flexibility - the ability to add new products in the system; process flexibility - the set of products/services that the enterprise can provide. Wadhwa and Rao [19] who emphasize flexibility as ability of implementation various scenarios, which help in undertaking the right decision. De Groote [10] perceive flexibility as a set of physical abilities corresponding to technological options. Perminova [5] suggest achieving synergy by using lessons learned simultaneously with flexibility and quick decision making in a reply to an investigated case. Authors in article define flexibility as capacity of adjustment to the market in short-term and long-term horizon of development, increasing NPV of enterprise, client satisfaction and lowering environmental impact. Flexible design should be divided into three main phases:

- Recognision of uncertainity and its characterization
- Simulation of performance for range of scenarios
- Evaluation (multidimensional)

Such a point of view and approach lead to reduction or elimination of downside risks and maximization of upside opportunities, which are unpredictable and often temporary caused by trend breakers. The most important trend breakers (unpredictable changeable events influencing disturbing linear forecast) are: inventions, politics (regulations, wars), economic crisis and booms, changes in social trends. Flexibility is the cheapest way to fulfill market demand.

2.4. Examples

Flexibility has been successfully applied in many different fields in civil engineering. The most spectacular and broadly described can be named: Ponte 25 de Abril in Lisbon Portugal, Vertical Expansion of Health Care Service Corporation Building, Chicago [21], off-shore oil platform investigated by Lin [12], buffer parking – Pasławski, Różdżyńska [11]. It is also applied in road infrastructure by preparation all devices for future expansion. All cases show increase of NPV, reduction of initial cost and risk. Very often there is no possibility to build structure in one phase mainly due to financial problems and time shortage. In such a cases flexible approach is inevitable. Flexibility in construction is successfully applied in temporary structures. An example of temporary expansion of stadium in London for Olympic Games in 2012. The stadium capacity was increased for the event and subsequently due to lower capacity demand platform was dismantled decreasing maintenance cost.

3. HOTEL BRANCH

Key factor of successful tourism industry is the ability to recognize and deal with changes of main trends in this branch [13]. There is a need for modification of business strategies to adopt to new opportunities and risks [7]. When some sort of problems appear in management of the hotel, enterprise can suffer the occurrence of 'strategic drift'. This phenomenon can be found when company gradually missing target and goes to unclear destination lowering its competiveness [15]. Cooperation between enterprise and customer as basis for strategic management requires three general indicators as in Fig. 3 [13].

3.1 Adjustment

Hospitality branch is burden with very high diversity and thus uncertainty. Main factors are: localization of the hotel, standard of service and many others. Hotels located in big cities have predictable market and precise type of client- due to stable conditions and relatively high demand. Resorts located in some distance form agglomerations are subjected to big amplitude of demand and aspects listed below, which have significant influence over financial fluency of enterprise.

- Number of visitors
- Variety of clients
- Length of stay (days, weeks, month, seasons)
- Seasonality
- Weather dependence
- Regional business dependence

There is a need to design infrastructure propitious to quick, simple and cheap adjustments to changing needs. Adjustments concern mainly size of house, access for disabled people, rearrangement for short and long term stay, heating system of higher capacity for future development of heating grid. Creation of flexible adjustment loop enables transformation to current demand save time and money increasing in the same time satisfaction of the client Fig. 4. Presents comparison of traditional and flexible approach to the demand fluctuation. Main disadvantage of traditional development is one-way development and is burdened with high risk in case of market demand change. Reduction threads can be achieved by multidirectional ability of service performance.

3.2 Case study

Paper investigates the case of small-scale holiday resort situated by the lake in vicinity of two bigger agglomerations in Poland. Resort belongs to private investor and has tradition of 20 years. Starting from one summerhouse gradually growing achieved level of 52 beds and hotel restaurant.

On the brink of development entrepreneur is about to take a decision on how to progress. As seen in Fig. 5 a) given historical data reveal irregular pattern showing periods of decrease of occupancy in years 2009-2011, growth in 2012, fall in 2013 and sharp increase in 2014. Similar tendency is shown by Fig. 5 b). Clearly visible is influence of world economic crisis. History of decision making in development of accommodation places (Fig. 5 c)) lead to conclusion, that building additional places have not been made adequately to market changes. After reaching high level of occupancy in 2009, decision of building 8 additional housing places has been made. Year after development occupancy dropped and generated insignificant increase in total sale incomparable to suffered costs. In long term additional houses were fully used three years later in 2013 and 2014 when number of peaks and occupancy have risen.

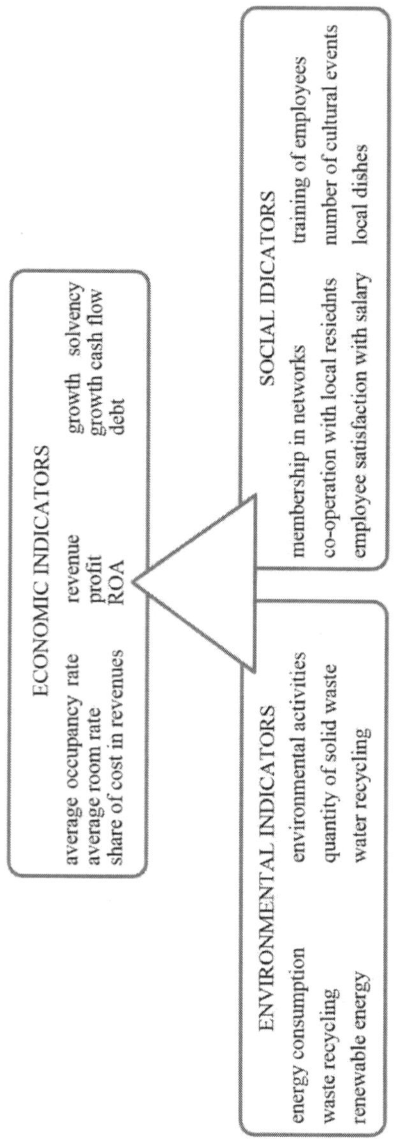

Figure 3. Economic, environmental and social indicators.[13]

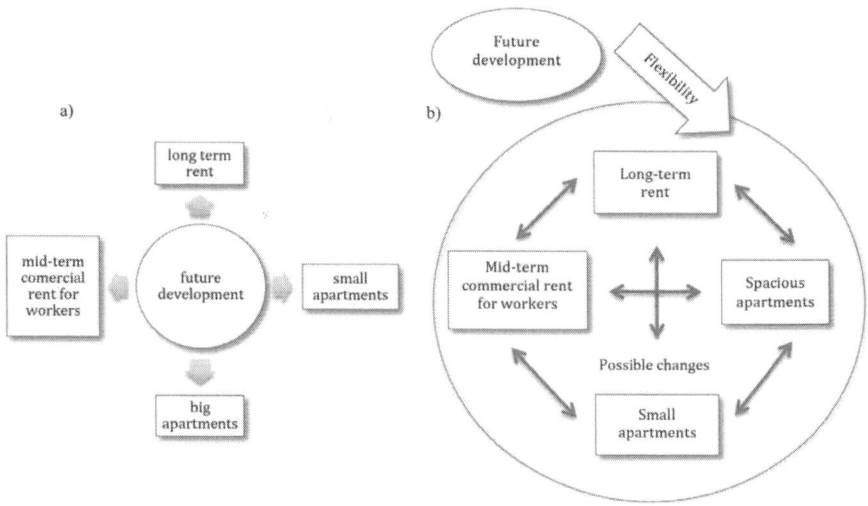

Figure 4. (a) Traditional design; (b) flexible adjustment.

Decision of future expansion is loaded by short and long-term trends namely: sales of 2 bed rooms percentage increases each year; short-term stay (weekend and holiday) are more preferable during off season term; long-term e.g. more than 6 months are more frequent each year. Moreover there are different reason for renting such as:

holidays, delegation or temporary apartment, thus differences in expectation appear e.g. various size of room is needed. Season time is predictable but off-season in strongly dependent on local investment and can oscillate substantially, moreover is difficult to predict. Proposed solution is based on constructing semidetached studios enabling temporary transformation into one apartment for different purpose fulfilling the demand for different kind of service following the idea presented by Fig. 4 b). Due to increasing number of factors needed to be taken into consideration future predictions seem heavily impacted by risk on the verge of uncertainty, what proofs urgent need of flexible management [16, 18].

In response to investors need for planning progress of his recreation center some prognosis of market growth have been made. Traditional method of forecasting revealed most probable scenario representing future. Nonetheless reality may become extremely different to most probable forecast, as experienced in 2009 during economic crisis, thus flexible methods of planning were taken into account with possible optimistic and pessimistic market growth. Those additional scenarios give wide horizon of possibilities to take into consideration. To foresee not only one solution, random number sampling method was used to create spectrum of possible market change presented in Fig. 6 a) [1].

Making case studies more evident, some assumptions have been made. Prognosis time is set to six years 2015- 2020. Investor decides to build 5 houses with 2 additional accommodation places each. Calculations [15,17] are made in Polish Złoty (1€ = 4,12 zł - NBP rate from 01.06.2015). Standard approach is represented by realization investment as a whole at once in year 2015 assuming, according to most probable scenario, rising tendency of market. Flexible approach is represented by construction of 1 additional house each year when market grows and not building at all when market falls.

To represent flexible approach idea 5 scenarios were chosen (Fig. 6 b)) as an examples to proceed with simulation:

- Optimistic – biggest possible market growth, sales goes up
- Most probable – normal market growth, sales goes up
- Pessimistic – market drop, sales goes down
- Optimistic goes pessimistic – biggest possible market growth at first, then sudden drop
- Pessimistic goes optimistic – market drop at first, then sudden growth

Optimistic scenario will preferably bring more income from traditional approach by exploiting opportunities given by additional accommodation places in times of great market growth. Flexible approach in this case, although it harnesses opportunities, due to moderate decision making model in effect brings somewhat less profit.

In pessimistic scenario building additional accommodation places contributes to suffering even greater losses than ones caused by hostile economical condition. When number of clients falls, houses stay empty generating no profit only maintenance costs, investment becomes unprofitable. Flexible solution can minimize losses caused by need of maintaining empty houses and prevent from spending savings on unnecessary accommodation.

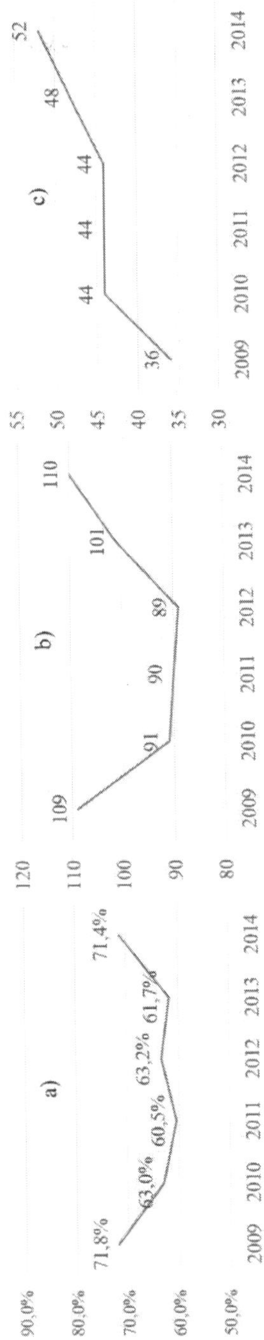

Figure 5. (a) Occupation - average annual; (b) accommodation peaks - days with over 90% occupancy; (c) accommodation places.

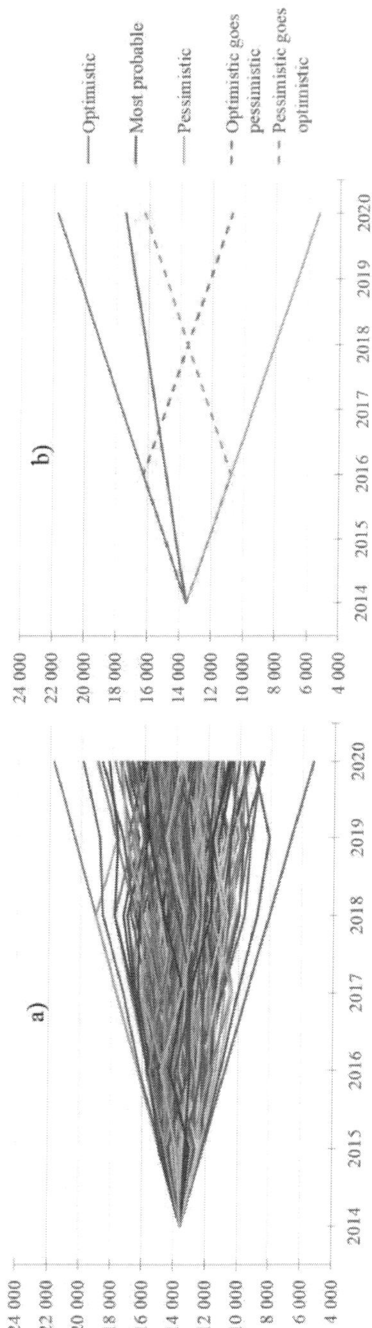

Figure 6. (a) Spectrum of probable market change; (b) chosen example scenarios.

Table 1. Calculations for most probable scenario

		2015	2016	2017	2018	2019	2020
Standard approach	Market demand	14 204	14 862	15 520	16 178	16 837	17 495
	Additional accomodation	10	0	0	0	0	0
	Income	498 187 zł	521 274 zł	544 361 zł	567 448 zł	590 536 zł	612 335 zł
	Building cost	-200 000 zł	0 zł	0 zł	0 zł	0 zł	0 zł
	Maintenance	-287 447 zł	-295 363 zł	-303 278 zł	-311 194 zł	-319 109 zł	-326 584 zł
	Profit	10 740 zł	225 911 zł	241 083 zł	256 255 zł	271 426 zł	285 752 zł
	Cumulative NPV	**10 228 zł**	**215 137 zł**	**423 393 zł**	**634 215 zł**	**846 884 zł**	**1 060 117 zł**
Flexible approach	Additional accomodation	2	2	2	2	2	0
	Income	498 187 zł	521 274 zł	544 361 zł	567 448 zł	590 536 zł	612 335 zł
	Building cost	-40 000 zł	-40 000 zł	-40 000 zł	-40 000 zł	-40 000 zł	0 zł
	Maintenance	-281 687 zł	-291 043 zł	-300 398 zł	-309 754 zł	-319 109 zł	-326 584 zł
	Profit	176 500 zł	190 231 zł	203 963 zł	217 695 zł	231 426 zł	285 752 zł
	Cumulative NPV	**168 095 zł**	**340 641 zł**	**516 832 zł**	**695 930 zł**	**877 258 zł**	**1 090 491 zł**

Due to specific relations of data most probable scenario shows insignificant benefit of flexible variant (Table 1.). This situation is caused by high share of building and maintenance costs and influence of interest rate in net present value method of calculation. Although knowing that in most probable scenario building additional houses in yearly manner is as profitable as building all of them at once, shows that in this particular case flexible solution is preferred

Scenarios optimistic goes pessimistic shown in Table 2 and pessimistic goes optimistic visible in Table 3 present extreme situations with drastic market change. Regardless of market changes flexible approach harnesses occasion out of steadily growing market protecting from suffering loss when tendency diverts. Both examples show superiority of flexible approach over traditional planning method giving adequately 139 398 zł and 126 258 zł greater net present value score.

5. CONCLUSIONS

Implementation of flexibility in shown example is not limited to housing or holiday market branch. Many companies may use flexibility as a mean of risk reducing and balanced form of development. Discussed example of flexibility was limited to showing decision making basing on economic calculations and number of housing places. Future studies may be conducted showing other possibilities of implementation flexibility on different levels such as diverting renting space, changing construction of houses, using different materials or energy sources.

Traditional methods of planning based on forecasts due to frequent market changes may become very inaccurate. Some economic factors are unpredictable so decision making is transferred to uncertainty background. Standard approach is associated with passive investing by enforcing one decision and anticipating future. Flexible approach means proactive investing, which enables investor to mitigate risk by stopping expansion when not necessary or growing when possible and harnessing occasions. As seen in the case study, proactive attitude gives great advantage no matter the economy changes.

Table 2. Calculations for optimistic goes pessimistic scenario.

		2015	2016	2017	2018	2019	2020
Standard approach	Market demand	14 917	16 288	14 917	13 545	12 174	10 803
	Additional accomodation	10	0	0	0	0	0
	Income	523 190 zł	571 281 zł	523 190 zł	475 100 zł	427 009 zł	378 918 zł
	Building cost	-200 000 zł	0 zł	0 zł	0 zł	0 zł	0 zł
	Maintenance	-296 019 zł	-312 508 zł	-296 019 zł	-279 531 zł	-263 043 zł	-246 555 zł
	Profit	27 171 zł	258 773 zł	227 171 zł	195 568 zł	163 966 zł	132 364 zł
	Cumulative NPV	**25 877 zł**	**260 591 zł**	**456 830 zł**	**617 724 zł**	**746 196 zł**	**844 968 zł**
Flexible approach	Additional accomodation	2	2	0	0	0	0
	Income	523 190 zł	571 281 zł	523 190 zł	475 100 zł	427 009 zł	378 918 zł
	Building cost	-40 000 zł	-40 000 zł	0 zł	0 zł	0 zł	0 zł
	Maintenance	-290 259 zł	-308 188 zł	-291 699 zł	-275 211 zł	-258 723 zł	-242 235 zł
	Profit	192 931 zł	223 093 zł	231 491 zł	199 888 zł	168 286 zł	136 684 zł
	Cumulative NPV	**183 743 zł**	**386 095 zł**	**586 066 zł**	**750 514 zł**	**882 371 zł**	**984 366 zł**

Table 3. Calculations for pessimistic goes optimistic scenario.

		2015	2016	2017	2018	2019	2020
Standard approach	Market demand	12 174	10 803	12 174	13 545	14 917	16 288
	Additional accomodation	10	0	0	0	0	0
	Income	427 009,0 zł	378 918,5 zł	427 009,0 zł	475 099,5 zł	523 190,1 zł	571 280,6 zł
	Building cost	-200 000,0 zł	0,0 zł	0,0 zł	0,0 zł	0,0 zł	0,0 zł
	Maintenance	-263 043,09 zł	-246 554,91 zł	-263 043,09 zł	-279 531,27 zł	-296 019,45 zł	-312 507,63 zł
	Profit	-36 034,08 zł	132 363,57 zł	163 965,92 zł	195 568,27 zł	227 170,62 zł	258 772,96 zł
	Cumulative NPV	**-34 318,17 zł**	**85 739,49 zł**	**227 379,41 zł**	**388 273,91 zł**	**566 268,03 zł**	**759 368,40 zł**
Flexible approach	Additional accomodation	0	0	0	2	2	2
	Income	427 009,0 zł	378 918,5 zł	427 009,0 zł	475 099,5 zł	523 190,1 zł	571 280,6 zł
	Building cost	0,0 zł	0,0 zł	0,0 zł	-40 000,0 zł	-40 000,0 zł	-40 000,0 zł
	Maintenance	-255 843,09 zł	-239 354,91 zł	-255 843,09 zł	-273 771,27 zł	-291 699,45 zł	-309 627,63 zł
	Profit	171 165,92 zł	139 563,57 zł	171 165,92 zł	161 328,27 zł	191 490,62 zł	221 652,96 zł
	Cumulative NPV	**163 015,16 zł**	**289 603,43 zł**	**437 462,99 zł**	**570 188,16 zł**	**720 226,06 zł**	**885 626,92 zł**

REFERENCES

1. S. Brandt, Analiza danych. Wydawnictwo Naukowe, Warszawa, 2002.
2. E. Osipova, E. Eriksson. Balancing control and flexibility in joint risk management: Lessons learned from two construction projects International Journal of Project Management 31 (2013) 391–399.
3. H. Volberda, Building the flexible firm. How to remain competitive. Oxford University Press, 1998.
4. J. Browne, Classification of flexible manufacturing systems, The FMS Magazine 04. (1984), 114-117.
5. O. Perminova, M. Gustafsson, K. Wikstrom, Defining uncertainty in projects – a new perspective, International Journal of Project Management 26 (2008) 73-79.
6. G. Johnson, K. Scholes, Exploring corporate strategy (4th ed.). London, 1997.
7. L. DwyeR, L. Knezevic Cvelbar, D. Edwards, T. Mihalic, Fashioning a destination tourism future: the case of Slovenia, Tourism Management 33 (2012) 305-316.
8. R. de Neufville, S. Scholtes, Flexibility in engineering design, Massachusetts Institute of Technology, 2011.
9. A.K. Sethi, and S.P. Sethi, Flexibility in Manufacturing: A survey, The International Journal of Flexible Manufacturing Systems 2 (1990) 289-328.
10. X. De Groote, The flexibility of production process: a general framework, Management Science 40 (1994) 933-945.
11. J. Paslawski, M. Rozdzynska, Flexible Approach in Infrastructure Design Buffer Parking Case Study. Procedia Engineering 57 (2013) 882 – 888.

12. J. Lin, Exploring Flexible strategies in engineering systems using screening models: applications to offshore petroleum projects, Cambridge, MA: PhD Dissertation, MIT Engineering Systems Division, 2009.

13. L. Knezevic Cvelbara, L. Dwyer, An importance-performance analysis of sustainability factors for long-term strategy planning in Slovenian hotels, Journal of Sustainable Tourism, 21(3) (2013) 487-504.

14. K. Das, Integrating effective flexibility measures into a strategic supply chain planning model, European Journal of Operational Research 211(1) (2011) 170-183.

15. H. Gawron, Opłacalność inwestowania na rynku nieruchomości, Wydawnictwo Akademii Ekonomicznej w Poznaniu, Poznań, 2006.

16. C.L. Pritchard, Risk Management Concepts and Guidance, Warszawa: WIG-Press., Warszawa, 2002.

17. E. Ostrowska, Ryzyko projektów inwestycyjnych, Polskie Wydawnictwo Ekonomiczne, Warszawa, 2002.

18. H. Henzel, Strategie inwestowania na rynku nieruchomości, Wydawnictwo Akademii Ekonomicznej im. Karola Adamieckiego w Katowicach, Katowice, 2009.

19. S. Wadhwa, K.S. Rao, A unified framework for manufacturing and supply chain flexibility, Global Journal of Flexible Systems Management 5 (2004) 15-22.

20. M. Lynch, U.S. Department of Energy.

21. Guma, J. Pearson, K. Wittels, R. de Neufville, D. Geltner, Vertical phasing as a corporate real estate strategy and development option, Journal of Corporate Real Estate, 11 (3) (2009) 144 - 157.

4

FACTORS OF THE EMPLOYMENT BEHAVIORAL INTENTION OF LEISURE AND HOSPITALITY MANAGEMENT COLLEGE STUDENTS IN TAIWAN

Jin-Li Hu*

Institute of Business and Management, National Chiao Tung University, Taiwan

ABSTRACT

This study examined the factors of the employment behavioral intention of leisure and hospitality management college students in Taiwan. The research framework was based on a literature review and relevant research variables. This study added an additional construct (perceived behavioral control) and a moderator (individual difference) into the theory of reasoned action (TRA) model. The subjects were college students in Taiwan. Questionnaires were administered and a total of 450 valid samples were collected. This study adopted two-step structural equation modeling (SEM), and both SAS and AMOS were adopted as the tools for analyzing the data for reconfirmation. A conceptual model was then proposed, in which the employment behavioral intention was influenced by positive internship experiences and negative internship experiences. Moreover, the relationship between positive internship experiences and employment behavioral intention was moderated by the internship system. A moderating test revealed that the influence of internship experiences on employment behavioral intention was stronger for leisure and hospitality management students working shorter hours than for students working longer hours.

KEYWORDS

Internship system; Attitude; Employment behavioral intention; Leisure and hospitality management

1. INTRODUCTION

Over the past decade, the total number of colleges and universities in Taiwan has increased every year. According to the statistical information from the Census and Statistics Department of the Ministry of Education, the number of colleges and universities increased from 155 in the academic year of 2001 to 163 in the academic year of 2010. Moreover, the number of universities nearly doubled, from 57 to 112, and the number of universities (institutes) of science and technology increased from 67 to 77. In addition, the number of students in colleges and universities increased from 1,084,012 in the academic year of 2001 to 1,124,425 in the 9 academic year of 2010, which was a growth of nearly 4% (National Statistics, Taiwan).

The above statistics indicated that it is a popular trend to pursue college or university education. However, according to the national statistics of Taiwan, the unemployment rate of individuals with a college degree or above from 2007 to 2011 was 4.51%, 4.78%, 5.98%, 5.62% and 5.18%, respectively, which was higher than the average unemployment rate. Moreover, according to the statistical information of the Directorate-General of Budget, Accounting and Statistics of the Executive Yuan (National Statistics, Taiwan) the youth unemployment rate was higher than that in neighboring Asian countries, such as Japan and South Korea. Therefore, the Ministry of Education has begun promoting school internship programs as an important policy to enhance the employability of college graduates. The National Youth Commission and individual colleges have also devoted themselves to similar programs. For example, the Ministry of Education has set industry-academy cooperation information to specially deal with internship programs, in order to connect schools and industries.

According to a survey of college youth employability investigated by the National Youth Commission in 2009 (National Youth Commission, Executive Yuan, The 2009 Survey on the employability of College Students), up to 67.6% of college students believed that internship experiences could effectively cultivate employability, followed by professional licenses and part-time jobs. In addition, the students believed that internship experiences could enhance their stability, resistance to stress, working attitude and communication skills. A good working attitude, stability, resistance to stress, and communication skills are the traits that are valued by employers. The results of previous showed that internship experiences during study periods can enhance employability to a certain extent. The main purpose in this study is to explore how internship experiences affect youth employability and whether internship experiences would significantly enhance youth employability or not. In addition, this study attempted to find if there are other mediators or moderators in this process, and confirm the value of internship experiences.

2. LITERATURE REVIEW

2.1 Internship

According to the Industry-Academy Cooperation Program implemented by the Ministry of Education, there are five types of internship programs: 1) summer courses; 2) semester courses; 3) academic year courses; 4) healthcare-related department courses; and 5) overseas internships (Industry-Academy-Cooperation-Information). The purpose of an internship is to enable students to integrate academic theory and practical work in order to understand the actual operational status of each department, confirm their learning, enhance their professional knowledge, and develop a correct understanding of workplace and professional ethics, which are all necessary job-finding skills.

After internship experiences, students can shorten the time required to explore the atmosphere of the workplace, and understand their own SWOT, learning lacking capacity to develop employability. Moreover, they can explore their future career development direction through internship programs.

Internships are not only useful to students. For parents, internship programs can set their minds at ease, as the host organizations are chosen by the school in order remove unsafe and unsound environments. Moreover, after the internship experience, students will become more mature and appreciate the hard work of their parents.

For schools, internship programs can help to understand the industry's current situation and enable them to adjust teaching styles to closely integrate with the enterprise environment. On the budget side, programs can save some money, as they may not have enough money to purchase the modern equipment that is used in the industries. In sum, internship programs can help schools become more efficient in both teaching and budget usage.

Internship programs also benefit the industry. Cooperative firms can cultivate the required human capital early, thereby reducing staff training costs and the time needed to find talented people. The schools' huge R&D power and human resources can also become the solid backing of the industry technology research and development. All in all, internship programs benefit each of the parties involved, including students, parents and industry.

Cannon et al. [1] stated that students are increasingly viewing internships as being less as a vehicle for augmenting their education, and more as a means of gaining a competitive edge in the marketplace for new jobs.

Brooks et al. [2] indicated that internship experiences, whether alone or in combination with previous work experience, are associated with higher levels of self-concept crystallization. In addition, the job characteristics of feedback, task variety, and opportunities to deal with others are positively related to changes in several career development constructs.

Knouse et al. [3] found that college internships offer a variety of benefits to students, both for improving performance while in college and for increasing opportunities for finding a job upon graduation. While students are still in college, internships can help them develop immediate skills that can improve their course performance, such as better time management, better communication skills, better selfdiscipline, heightened initiative and an overall better self-concept. In addition, internships allow students to directly access job sources, impress potential employers, build confidence in their job search, hone their work values, and build social skills that are beneficial during employment interviews.

Margolis et al. [4] concluded that five factors might influence postgraduation attrition: 1) feeling prepared to pursue an engineering career; 2) internship experiences; 3) senior capstone project course experiences; 4) satisfaction with the quality of instruction in engineering programs; and 5) career values related to financial rewards and enjoying co-workers.

However, there is literature that holds a different view. Aggett et al. [5] proposed that although students are expected to take the opportunity to enhance their competitive advantages via an internship, this was not the case (Table 1). Their study was aimed at hospitality, tourism and events management undergraduates at a British university. According to the data, the number of students who had undertaken an internship had progressively decreased since 2007. The student's reasons for being unwilling to take part in an internship included already having acquired industry experience, a desire to complete their degrees sooner, a contact having already been made in the industry, and no longer wanting to work in the industry.

Many student lack an understanding of the value of work experience and determination, and this precondition will decide whether the internship experience will provide a benefit or not.

2.2 Theory of Planned Behavior

The theory of reasoned action (TRA), developed by Fishbein et al. [6] is an attempt to explain volitional behaviors.

Ajzen extended TRA into the theory of planned behavior (TPB) in 1985. Ajzen [7] concluded that the attitude toward the behavior, the subjective norm with respect to the behavior, and the perceived control over the behavior can usually predict behavioral intention with a high degree of accuracy. These intentions, in combination with the perceived behavioral control, can account for a considerable proportion of variance in behavior.

Attitude toward the behavior refers to the degree in which a person has a favorable or unfavorable evaluation of the behavior in question. Attitudes are more important when people are action-oriented. Concepts referring to behavioral dispositions, such as social attitude and personality trait, have played an important role in attempts to predict and explain human behavior [8-10]. However, attitude does not refer to the attitude towards general things, but the evaluation of an individual on positive and negative engagements in certain behaviors [6].

Subjective norm refers to the perceived social pressure to perform or not to perform the behavior. It is concerned with the likelihood that the specific individuals or groups (referents) with whom the individual is motivated to comply will approve or disapprove of the behavior [11].

Perceived behavioral control refers to people's perception of the ease or difficulty of performing the behavior of interest, or other factors defined as the perceived probability of succeeding at a given task [7]. Zhang et al. [12] assessed individuals' perceived competence to respond to novel or difficult situations and to deal with any associated obstacles or setbacks. Perceived behavioral control is a key variable that refers to beliefs about access to resources and the opportunities needed to perform a behavior [13].

Behavioral intention refers to the inclination to take action in the future based on personal subjective judgments, which can be used to predict the behaviors of an individual. Behavioral intention can find expression in behavior only if the behavior in question is under volitional control, i.e., if the person can decide at will to perform or not perform the behavior [6].

Finally, intentions are assumed to capture the motivational factors that influence a behavior. They are indications of how hard people are willing to try and how much of an effort they are planning to exert, in order to perform the behavior.

Table 1: Total Percentage of Tourism, Hospitality and Events Students Being Placed Per Year.

Academic Year	2005/06	2006/07	2007/08	2008/09	2009/10	2010/11
% of students being placed	25.9	27.2	37.2	26.7	12.9	10.4

Source: Aggett Mandy & Busby Graham (2011)

2.3 Experience

Fazio et al. [14,15] and Regan et al. [16] provided support for the notion that the attitudes formed through direct behavioral experience with the attitude object are more predictive of later behavior toward that object than are attitudes based upon indirect, non-behavioral experience. In other words, attitudes come from direct behavioral experience.

Table 2: Criteria of Goodness of Fit.

	Chi-square	GFI	AGFI	SRMR	RMSEA	CFI	NFI	NNFI
Criteria model	<3	≥0.9	≥0.8	≥0.1	≤0.08	≥0.9	≥0.9	≥0.9

Table 3: Percentage of Individual Differences.

Traits	Items	Percentage
School system	University of technology	83.6%
	General college	16.4%
Year of study during the internship	Freshman	0%
	Sophomore	1%
	Junior	55.4%
	Senior	43.6%
Gender	Male	54.7%
	Female	45.3%
Age	18 years old	0.0%
	19 years old	0.6%
	20 years old	6.5%
	21 years old	44.0%
	22 years old	35.6%
	23 years old	10.5%
	Over 24	2.8%
Pre-admission educational background	Graduate from department of leisure management	30.7%
	Graduate from department of hospitality management	17.3%
	Graduate from non-leisure-related departments	52%

Working experience before the internship	None Less than three months More than three months and less than six months More than six months and less than one year More than one year	45.8% 12.7% 10.8% 15.2% 15.5%
Location of the internship	Northern Taiwan Central Taiwan Southern Taiwan Eastern Taiwan Off-island	29.1% 23.8% 46.1% 0.3% 0.6%

Table 4: Percentage of Individual Differences (continued).

Traits	Items	Percentage
Type of internship unit	Hotel industry	14.2%
	F&B industry	22.6%
	Travel agency	5.3%
	Airline company	0.0%
	Leisure club	27.9%
	Leisure farm	4.0%
	National park	0.9%
	Theme park	5.6%
	Other	19.5%
Internship system	Below 200 hours	0%
	Between 201~ 500 hours	0%
	Six months	51.7%
	One year	48.3%
	Other	0%
Type of job	Rotational job	38.7%
	Fixed job	59.8%
	Other	1.5%

The classification of experiences can be divided into the five aspects of affective experiences, sensory experiences, physical experiences, relational experiences, and creative cognitive experiences [17].

Affective experiences are the emotions of individuals. Sensory experiences refer to perceptual experiences from the senses of sight, hearing, touch, taste and smell. Physical experiences are received by taking action and interacting with others. Creative cognitive experiences refer to cognitive experiences from rational thinking, and relational experiences are the module that contains aspects of the above four experiences.

The attitudes formed through direct behavioral experience with the attitude object are more predictive of later behavior toward that object than are attitudes based upon indirect, non-behavioral experiences [18].

Table 5: Analysis on the Extent of Agreement on the Factors of Positive Experiences

Aspects of factors of positive experiences	Mean	SD
Factor 1: Affective experiences		
The internship enabled me to make many friends, expanded my vision, and made me feel that the effort was worth it.	4.06	0.81
The memories that the internship gave me made me feel that the hard work was worth it.	3.95	0.80
The results of the internship made me feel that it was worth it to spend the time and effort.	3.80	0.84
As a whole, I was interested in the internship.	3.78	0.82
As a whole, the internship made me feel happy.	3.75	0.79
Factor 2: Sensory experiences		
The affirmation of customers from the internship made me feel a strong sense of accomplishment.	4.11	0.77
The internship cultivated my spirit of service for the leisure industry.	3.87	0.71
The internship made me feel satisfied because I could apply what I had learnt.	3.75	0.85
I believe that the internship can help me obtain employment.	3.74	0.80
The internship aroused my interest in the expertise of the leisure industry.	3.60	0.80
The internship enabled me to apply the theories that I had learnt at school.	3.42	0.87
Factor 3: Physical experiences		
The internship made me face the reality of society.	4.20	0.75
The internship increased my social experiences.	4.20	0.77
The internship increased my tolerance and patience for the management of tasks and work.	4.03	0.74
The internship helped me improve my personal words and deeds.	3.78	0.75
The internship changed my original living habits.	3.60	0.87

Factor 3: Physical experiences		
The internship made me face the reality of society.	4.20	0.75
The internship increased my social experiences.	4.20	0.77
The internship increased my tolerance and patience for the management of tasks and work.	4.03	0.74
The internship helped me improve my personal words and deeds.	3.78	0.75
The internship changed my original living habits.	3.60	0.87
Factor 4: Relational experiences		
The internship gave me the desire to influence the people around me to improve their level of understanding of the leisure culture.	3.61	0.77
The internship gave me the desire to change the viewpoints of those who had a negative opinion of the leisure industry.	3.57	0.82
The internship gave me the desire to think like the employers in the leisure industry regarding how to improve the overall quality of the leisure industry.	3.56	0.74
Factor 5: Creative cognitive experiences		
The internship helped me to reflect on my career planning.	3.94	0.78
The internship inspired me to strengthen my own abilities and improve my own values.	3.90	0.76
The internship aroused my intention to live a better life.	3.88	0.79
The internship made me think whether I should continue staying in leisure industry to obtain employment.	3.69	0.82
The internship aroused my ambition to gain an in-depth understanding of knowledge relevant to the leisure industry.	3.51	0.82

2.4 Attitude-Behavior Relation

Individual difference variables that have been found to moderate attitude-behavior correspondence include the degree of affectivecognitive consistency [19]. Similarly, a number of situational variables appear to moderate the attitude-behavior relation.

For example, Warner et al. [20] found that the public versus private nature of behavior interacts with attitude in influencing attitude-behavior consistency. Snyder et al. [21] demonstrated that situations that increase the relevance of salient attitudes as guides to actions also increase the correspondence between attitude and action. Fazio et al. [22] indicated that the manner of attitude formation and self-monitoring has been shown to moderate the attitude-behavior relation. Lin et al. [23] proposed that peer-reporting intentions are influenced simultaneously by the subjective norm, ethical judgments, and attitudes. Additionally, the relationships are moderated by a joint moderator, which combines ego strength and ethical climate.

Table 6: Analysis on the Extent of Agreement on the Factors of Negative Experiences.

Aspects of factors of negative experiences	**Mean**	**SD**
Factor 1: Sensory experiences		
The internship made me feel that I was low-cost labor, which made me feel unpleasant.	3.12	1.02
I felt unpleasant because my position as an intern made it possible to be ordered around by others.	3.06	0.95
I felt unpleasant because my position as an intern made it possible to be treated unequally.	2.96	0.94
The job content of the internship failed to make me learn more.	2.59	0.96
I could not adapt to the environment of the internship.	2.55	0.87
I could not adapt to the life of the internship.	2.52	0.90
I could not adapt to the work of the internship.	2.50	0.85
I could not bear the workload of the internship.	2.48	0.89
The internship made me feel that it was simply work, and that I could not learn knowledge or skills from it.	2.48	0.95
The internship made me have a negative opinion of the leisure industry.	2.44	0.96
To me, participating in the internship was simply for obtaining credits and meeting the requirements of the school.	2.38	1.02
The internship made me feel that it was a waste of time.	2.33	0.95
Factor 2: Physical experiences		
The internship changed my work values.	3.50	0.86
The internship made me disapprove of the payment and fringe benefits of the leisure industry.	3.07	0.98
The internship made me disapprove of the vacation system of the leisure industry.	2.92	0.88
The internship made me disapprove of the frequent overtime work of the leisure industry.	2.91	0.91
The internship made me disapprove of the work environments of the leisure industry.	2.70	0.84
Factor 3: Affective experiences		
As a whole, the internship made me feel nervous.	2.79	1.02
As a whole, internship made me feel stressed.	2.59	0.99
As a whole, the internship made me feel frustrated.	2.53	0.94
As a whole, the internship made me feel miserable.	2.49	1.00
Factor 4: Creative cognitive experiences		
The internship made me feel that my work ability could not meet the needs of the leisure industry.	3.27	0.97
The internship made me feel that I still lacked the ability to enter the workplace.	3.12	1.00
Factor 5: Relational experiences		
The internship made me learn the inveteracy of society.	2.52	1.02
The internship made me feel that I was not socialized for early exposure to society.	2.38	0.93

Table 7: Analysis on the Extent of Agreement on Employment behavioral intention.

Aspects of perceived behavioral control and employment behavioral intention	Mean	SD
Employment behavioral intention		
1. I was satisfied with the overall internship experience.	3.54	0.86
2. After experiencing the internship, I am still willing to return to the original internship unit for employment if there is an opportunity.	3.30	1.00
3. After experiencing the internship, I am still willing to engage in the leisure industry.	3.52	0.85
4. After experiencing the internship, I will recommend other people to engage in the leisure industry.	3.42	0.80
Perceived behavioral control		
1. After experiencing the internship, my internship unit caused me to have thoughts of changing my internship unit.	3.85	1.15
2. After experiencing the internship, I had thoughts of dropping out of school.	2.90	1.12
3. After experiencing the off-campus experiences, I would like to have a career transition (e.g. pursuing further education, switching my career to work as teaching staff, etc.)	2.89	1.01
4. When I faced a problem during the internship, I would complain to the supervisor or my colleagues.	3.19	0.99

In summary, several personal and situational factors have been found to affect the attitude-behavior relation [18]. The workings of the aggregation principle show that general attitudes and personality traits do in fact predict behavioral aggregate much better than they predict specific behaviors [8]. Therefore, according to the above literatures, this study assumed that the attitude toward internship experiences and the perceived behavioral control will affect employment behavioral intention. Moreover, individual differences will moderate the relation between attitude and employment behavioral intention (Table 12).

3. RESEARCH METHODOLOGY

3.1 Conceptual framework

Differing from the original model in Chen et al. [24], this study divided positive and negative attitudes toward internship experiences into two different constructs, added the internship system as a moderator, and tried to classify another independent variable (perceived behavioral control) from the original questionnaire in Chen et al. [24] (Figure 3).

Table 8: Standardized Loadings of the Indicators and Cronbach' α.

Construct	Cronbach's α	Measurement Item	Standardized loadings	(t-value)
Positive Experiences	0.89	X1	0.93	18.21
		X2	0.75	14.37
		X3	0.64	12.45
		X4	0.72	12.82
		X5	0.68	12.94
Negative Experiences	0.83	X6	0.81	16.88
		X7	0.90	19.79
		X8	0.64	12.25
		X9	0.66	12.98
		X10	0.37	6.56
Perceived behavioral control	0.71	P1	0.74	12.61
		P2	0.59	10.46
		P3	0.56	9.30
		P4	0.42	6.93
Employment behavioral intention	0.78	B1	0.77	14.68
		B2	0.75	15.06
		B3	0.55	10.10
		B4	0.48	8.49

Table 9: Discriminant Validity by the Chi-Square Difference Test.

Constructs	Unconstrained (d.f.=98)	Constrained (d.f.=99)	Chi-Square difference (d.f.=99-98)	Conclusion
F1 F2	142.62	430.67	288.05	Supported
F1 F3	142.62	267.99	125.37	Supported
F1 F4	142.62	262.32	119.70	Supported
F2 F3	142.62	194.99	52.37	Supported
F2 F4	142.62	322.03	179.41	Supported
F3 F4	142.62	267.42	124.80	Supported

Table 10: Goodness of Fit.

	Chi-square/df	GFI	NFI	NNFI	AGFI	SRMR	RMSEA	CFI
Model fitness	1.46	0.95	0.95	0.98	0.92	0.06	0.04	0.98
Criteria	<3	≥0.9	≥0.9	≥0.9	>0.8	<0.1	<0.08	≥0.9

Table 11: Path Analysis.

	Path coefficient	t-value	Result
PF1F4	0.61	4.59	Significant
PF2F4	-0.44	-2.35	Significant
PF3F4	-0.16	-0.90	Not significant

3.2 Research hypotheses

According the above conceptual framework, this study proposed the following hypotheses:

H1: Positive attitudes toward the internship experience have a positive influence on employment behavioral intention.

H2: Negative attitudes toward the internship experience have a negative influence on employment behavioral intention.

H3: The perceived behavioral control of the internship has a positive influence on employment behavioral intention.

H4: The internship system moderates the relation between positive attitudes and employment behavioral intention.

H5: The internship system moderates the relation between negative attitudes and employment behavioral intention (Tables 13 and 14) (Figure 4).

3.3 Operational Definition

There were four constructs in the conceptual model: positive attitude, negative attitude, perceived behavioral control and employment behavioral intention. Based on the theory of planned behavior [14-16], this study measured attitude through experiences. Moreover, this study categorized all experiences as being either positive or negative. Such experiences included affective experiences, sensory experiences, physical experiences, relational experiences, and creative cognitive experiences [19,24].

Table 12: The Moderation Effect of Positive Attitudes toward Internship Experience and Employment Behavioral Intention

Hypothesis Results for Moderating Effects (Positive attitude→employment behavioral intention)				
Hypothesis	Unconstrained (d.f.=217)	Constrained (d.f.=216)	Coefficients	Conclusion
X→Y	χ^2 =485.5	χ^2 =495.1 (9.6)	(A,B): (0.66, 0.48)	Supported

Note: A denotes 201~500 hours; B denotes six months

Table 13: The Moderation Effect of Negative Attitudes toward Internship Experience and Employment Behavioral Intention.

Hypothesis Results for Moderating Effects (Negative attitude→employment behavioral intention)				
Hypothesis	Unconstrained (d.f.=217)	Constrained (d.f.=216)	Coefficients	Conclusion
X→Y	χ^2= 485.5	χ^2 = 486.7 (1.2)	(A,B): (-0.60, -0.61)	Not supported

In Figure 3, the measurement item X1 denotes positive affective experiences, X2 denotes positive sensory experiences, X3 denotes positive physical experiences, X4 denotes positive relational experiences, X5 denotes positive creative cognitive experiences, X6 denotes negative affective experiences, X7 denotes negative sensory experiences, X8 denotes negative physical experiences, X9 denotes negative relational experiences, and X10 denotes negative creative cognitive experiences.

Table 14: Summary of Hypotheses Test.

Hypotheses	Results
H1: Positive attitudes toward the internship experience will have a positive influence on employment behavioral intention.	Supported
H2: Negative attitudes toward the internship experience will have a negative influence on employment behavioral intention.	Supported
H3: The perceived behavioral control of the internship will have a positive influence on employment behavioral intention.	Not supported
H4: The internship system will moderate the relation between positive attitudes and employment behavioral intention.	Supported
H5: The internship system will moderate the relation between negative attitudes and employment behavioral intention.	Not supported

Since the background theory was based on the theory of reasoned action and the theory of planned behavior, the rest of the constructs were perceived behavioral control and employment behavioral intention. This study reused the questionnaire of Chen et al. [24] to extend the previous findings; however, there were no items in the questionnaire that could be classified into the subjective norm construct (Figure 1).

According to Ajzen [7], Zhang et al. [12], and Kang et al. [13], this study classified P1~P4 into perceived behavior control, in which P1 refers to "After experiencing the internship, my internship unit caused me to have thoughts of changing my internship unit," P2 refers to "After experiencing the internship, I had thoughts of dropping out from school," P3 denotes "After experiencing the off-campus experiences, I would like to have a career transition," and P4 denotes "When I faced a problem during the internship, I would complain to the supervisor or my colleagues."

According to Fishbein et al. [6], employment behavioral intention is measured by Y1 to Y4, in which Y1 denotes "I was satisfied with the overall internship experience," Y2 denotes "After experiencing the internship, I am still willing to return to the original internship unit for employment if there is an opportunity," Y3 refers to "After experiencing the internship, I am still willing to engage in the leisure industry," and Y4 refers to "After experiencing the internship, I will recommend other people to engage in the leisure industry."

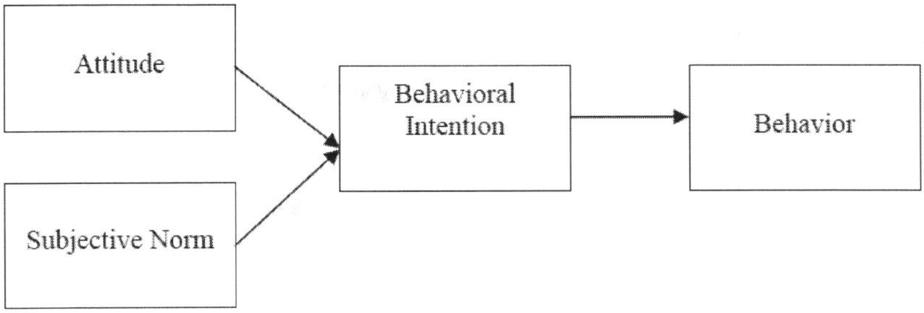

Figure 1. Theory of Reasoned Action (TRA).

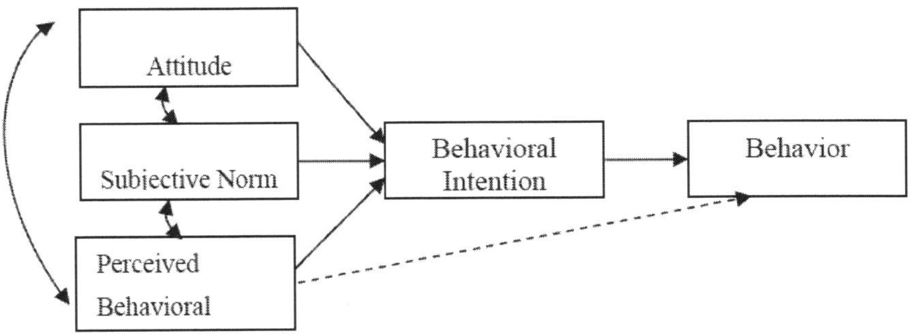

Figure 2. Theory of Planned Behavior.

Finally, based on Davidson et al. [18], Norman et al. [19], and Lin et al. [23], many kinds of individual differences can moderate the path between attitude and employment behavioral intention. Therefore, this study defined the internship system as a moderator.

3.4 Source of the Questionnaire

This study used the questionnaire of Chen et al. [24] in an attempt to extend the previous findings. Positive and negative attitudes were divided into two different constructs, adding one more independent variable and moderate variable to the model.

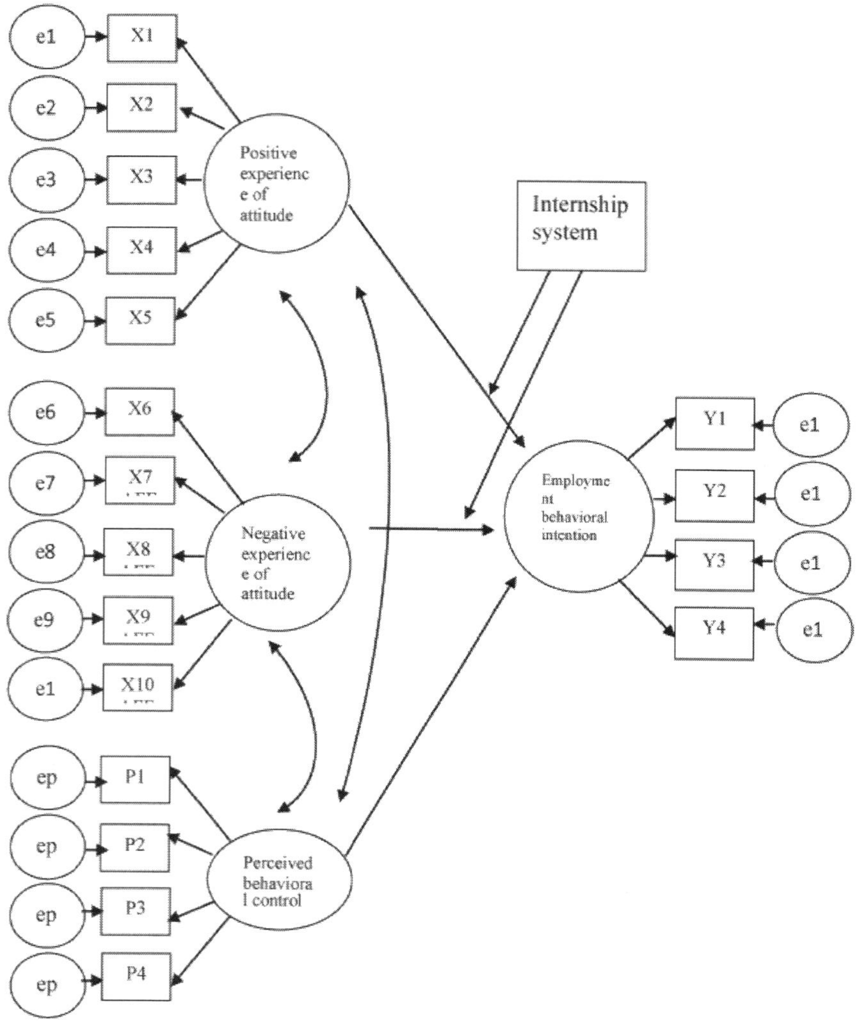

Figure 3. Conceptual Framework.

Chen et al. [24] collected data through a questionnaire survey. The participants were leisure and hospitality management college students in 2009 that had already completed an internship program. The participants came from four universities/colleges in northern, central and southern Taiwan, including Vanung University, National Taiwan University of Physical Education and Sport, Tajen University, and Yung Ta Institute of Technology and & Commerce. The questionnaire survey was conducted during October 2009. A total of 500 questionnaires were distributed and 490 were returned, showing a return rate of 98%. After eliminating 40 invalid questionnaires, there were a total of 450 valid questionnaires for analysis.

The contents of the questionnaire were categorized into four parts. The first part was positive experiences, which were divided into sensory experiences, affective experiences, creative cognitive experiences, physical experiences, and relational experiences. The second part was negative experiences, which were divided in the same way as positive experiences. The third part was employment behavioral intention and perceived behavioral control, and the final part was individual differences.

3.5 Analysis Methods

After data collection, SEM was applied for data analysis. Following the instructions of Anderson et al. [25], this study adopted two-step structural equation modeling (SEM). Both SAS and AMOS were adopted as the tools for analyzing the data for reconfirmation.

Descriptive statistics, such as percentage, mean, and standard deviation, were used to find out the distribution of the valid samples for each variable. Structural equation modeling (SEM) was then conducted to do confirmatory factor analysis (CFA), path analysis and goodness of fit. Finally, the chi-square difference test was adopted to analyze the moderation effect.

3.6 Descriptive statistics analysis

Regarding the valid returned samples, the frequency distributions and percentages of the data of students' individual differences were analyzed.

3.7 Reliability analysis

This study used Cronbach's α and composite reliability (CR) to measure the reliability. Nunnally [26] argued that Cronbach's α should greater than 0.7 (Table 8).

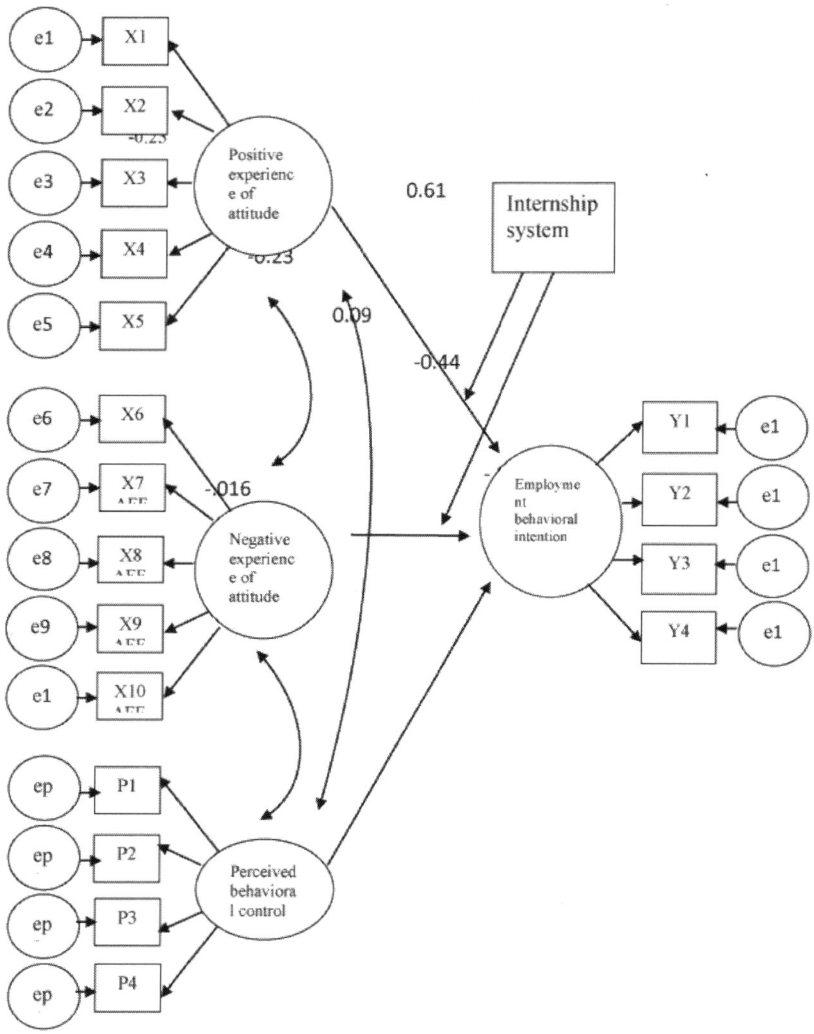

Figure 4. Hypotheses Testing of the Conceptual Model.

3.8 Validity analysis

The convergent validity and discriminant validity are usually recommended to measure validity. As for the convergent validity, composite reliability (CR) and average variance extracted (AVE) are usually used for analysis. This study compared the chi-square value under both the constrained model and the unconstrained model to test the discriminant validity.

3.9 Structural equation modeling analysis

In adopting structural equation modeling (SEM), the sample data was as suggested by the literatures:

(1) Bentler et al. [27]

If the data is normally distributed, there should be five samples of the items; otherwise, there should be ten samples..

(2) Anderson et al. [25]

The number of sample data should be at least 100~150.

(3) Hair et al. [28]

The sample data should be at least five times the estimators, and ten times is even better.

3.10 Goodness of fit

According to Hair et al. [28] the value of GFI, AGFI, NFI, and CFI should be above 0.9. A value of RMR and RMSEA that is less than 0.05 suggests a nice fit between the model and data [29]. A value of SRMR that is less than 0.08 indicates an acceptable model [30], and $\chi 2/d.f.$ is recommended to be below 3 [31]. Finally the value of NNFI is also recommended to be greater than 0.9 [32].

3.11 Chi-square difference test for the moderating effect

This study used the chi-square test to examine the moderation effect. Under the criteria of $\alpha=0.1$, if the chi-square difference between the unconstrained model and the constrained model is greater than 2.706, the moderating effect is significant. Under the criteria of $\alpha=0.05$, if the chi-square difference between the unconstrained model and the constrained model is greater than 3.84, the moderating effect is significant (Table 9).

3.12 Data Analysis

3.12.1 Descriptive statistics analysis

Regarding the valid returned samples, the percentages of the individual differences (Table 3) are analyzed. A total of 83.6% of the students came from universities of technology, and seniors accounted for 55.4% of the total, followed by juniors (43.7%). Females accounted for 54.7% and males accounted for 45.3% of the total. Most of the students were 21 or 22 years old, accounting for 46.2% and 33.8%, respectively. As for the pre-admission educational background, most of the students had non-leisure-related majors (52%), followed by leisure-related majors (30.7%). As for working experiences in leisure industries before the internship, 44% of the students did not have relevant experiences. Most of the internships were located in southern Taiwan (43.1%), followed by northern Taiwan (39.6%). Most of the internship units were food and beverage industry-related (28.9%), followed by leisure clubs (20.7%). Most of the internships lasted for six months (51.7%), followed by one year (48.3%). Finally, fixed jobs accounted for 62.2% of the total, followed by rotational jobs (36%). An overview of the total individual differences, positive experiences, negative experiences, perceived behavioral control and employment behavioral intention is as follows (Tables 4-7).

3.12.2 Structural equation modeling analysis

After moderating the goodness of fit, this study checked each item to see whether the absolute t-value exceeded 1.96 or not (under $\alpha=0.05$). If the absolute t-value was below 1.96, the convergent validity did not pass, the items would be deleted, and the analysis would be run again. The result showed that all t-values of the measurement items met the criteria (Table 10).

Regarding the discriminant validity, chi-square difference needed to exceed 3.841 under the criteria of $\alpha=0.05$. The result showed that all constructs met the criteria.

3.12.3 Goodness of fit

After adjusting, the GFI, NFI, NNFI, AGFI, SRMR, RMESA and CFI all passed the criteria. The result showed that this model had an acceptable goodness of fit (Table 2).

3.12.4 Path analysis

Under the criteria of $|t| \geq 1.96$ ($\alpha=0.05$), the path between positive and negative attitudes from experience and employment behavioral intention was significant. However, the path between perceived behavioral control and employment behavioral intention was not significant (Table 11).

3.12.5 Chi-square difference test for moderating effect

The questionnaire contained two individual difference items answered with the two categories of school system and internship system. School system referred to the regulations of school organization, the curriculum, learning, and age restrictions. The options were twoyear universities of technology, universities of technology, and general colleges. Of the total number of schools, 77% were universities of technology and 23% were general colleges.

Internship system referred to the working hours at the students' internship locations. The options were below 200 hours, 201-500 hours, six months, one year, and other. Of the total, 52% were 201-500 hours and 48% were six months.

The school system as not suitable as moderator, but it was suitable as a control variable. Therefore, the moderator in this study was the internship system. Under the criteria of $\alpha=0.1$, a chi-square difference greater than 2.706 indicated a significant moderating effect. As Table 12 shows, the moderating effect between positive attitudes toward the internship experience and employment behavioral intention was significantly supported. As Table 13 shows, the moderating effect between negative attitudes toward the internship experience and employment behavioral intention was not supported.

The result showed that more working time would lead to antipathy. The students whose working time was between 201-500 hours were more willing to continue being engaged in the tourism and leisure industry than the students whose working time was six months. As to why the moderating effect between negative attitudes toward the internship experience and employment behavioral intention was not significant, this study inferred that negative experiences would lead to disappointment and tiredness; therefore, a longer or shorter internship period would not matter. The hypotheses testing of the conceptual model was as follows, and a summary is shown in Table 13.

4. CONCLUSION

The study expanded the findings of Chen et al. [24] by separating the positive and negative attitudes toward internship experiences as well as by adding the independent variable of perceived behavioral control and a moderator to the structure model.

This study investigated the formation of attitudes formed through direct behavioral experiences. Perceived behavioral control refers to people's perception of the ease or difficulty of performing the behavior of interest, or other factors defined as the perceived probability of succeeding at a given task. Employment behavioral intention in this study referred to the willingness to engage in the leisure industry after experiencing an internship.

Just as Ajzen [7] indicated, the results showed that the behavioral intention to perform a behavior can be predicted with high accuracy from the attitudes toward the behavior and the perceived behavioral control. In this study, both positive and negative attitudes toward internship experiences had a significant effect on employment behavioral intention. As to why the perceived behavioral control of the internship did not have a significant influence on employment behavioral intention, the study inferred that the measurement items of perceived behavioral control were not direct enough.

In addition, a moderating effect was found to exist. The internship system will moderate the path between positive attitudes toward the internship experience and employment behavioral intention. The moderation result showed that students whose working time was between 201-500 hours were more willing to continue to engage in the tourism and leisure industry than were the students whose working time was six months. As to why the moderating effect between negative attitudes toward the internship experience and employment behavioral intention was not significant, this study believed that the attitude toward the internship experiences made students feel disappointment and tiredness, and that the length of the internship would therefore have no effect on their employment behavioral intention.

5. MANAGERIAL IMPLICATIONS

Attitudes about the experience will affect employment behavioral intention significantly. Among the five kinds of experiences, affective experiences and sensory experiences account for most of the experiences. Affective experiences are mainly feelings and emotions that touch an individual; sensory experiences impact the pattern of stimuli, processes and consequences, such that these are most of the sources of overall experience. Therefore, when promoting internship programs, students, schools and cooperative companies should pay attention to affective experiences and sensory experiences, so that the internship experience will be effective.

None of the literature has empirically investigated the moderating effect between the internship experiences and employment behavioral intentions of students majoring in the leisure-related fields. The result of this study showed that the students whose working period was between 201-500 hours had more willingness to continue to engage in the tourism and leisure industry after their internship experience.

It could be inferred that the students whose internship systems were shorter did not understand the tourism and leisure industry sufficiently. Therefore, they held more expectations for the hospitality and leisure industry than the students whose internship systems were longer. In addition, the more working hours they had, the more tiredness they would feel, while more working hours also brought with more understanding of the job. The results were perhaps related to job burnout. Overall, students whose working hours were six months had less willingness to engage in the tourism and leisure industry than were students whose working hours were below six months after experiencing the internship.

Future studies could discuss job burnout during internship experiences in the tourism and leisure industry to improve the phenomenon. Moreover, future studies could also discuss the real behavior and extent the model to complete the theory of planned behavior (Figure 2).

REFERENCES

1. Cannon JA, Arnold MJ (1998) Student expectations of collegiate internship programs in business: A 10-year update, Journal of Education for Business 73: 202-205.

2. Brooks L, Cornelius A, Greenfield E, Joseph R (1995) The relation of career-related work or internship experiences to the career development of college seniors, Journal of Vocational Behavior 46: 332-349.

3. Knouse SB, Tanner JT, Harris EW (1999) The relation of college internships, college performance and subsequent job opportunity, Journal of Employment Counseling 36: 35-43.

4. Margolis J, Kotys SV (2010) The post-graduation attrition of engineering students: an exploratory study on influential career choice factors, ASME International Mechanical Engineering Congress and Exposition 7: 449-446.

5. Aggett M, Busby G (2011) Opting out of internship: Perceptions of hospitality, tourism and events management undergraduates at a British University, Journal of Hospitality Leisure Sport & Tourism Education 10: 106-113

6. Fishbein M, Ajzen I (1975) Belief, Attitude, Intention, and Behavior: An Introduction to Theory and Research. Addison-Wesley, Reading, MA.

7. Ajzen I (1991)The theory of planned behavior, Organizational Behavior and Human Decision Processes 50: 179-211.

8. Ajzen I (1988) Attitudes, personality, and behavior, Chicago: Dorsey Press.

9. Campbell DT (1963) Social attitudes and other acquired behavioral dispositions. In Koch S (ed) Psychology: A study of a science.McGrawóHill, New York.

10. Sherman SJ, Fazio RH (1983) Parallels between attitudes and traits as predictors of behavior. Journal of Personality 51: 308-345.

11. Armitage CJ, Conner M (2001) Efficacy of the theory of planned behaviour: A meta-analytic review, British Journal of Social Psychology 40:471-499.

12. Zhang JX, Schwarzer R (1995) Measuring optimistic self-beliefs: A Chinese adaptation of the General Self-Efficacy Scale. Psychologia: An International Journal of Psychology in the Orient38: 174-181.

13. Kang H, Hahn M, Fortin DR, Hyun YJ, Eom Y (2006) Effects of Perceived Behavioral Control on the Consumer Usage Intention of E-coupons. Psychology & Marketing 23: 841-864.

14. Fazio RH, Zanna MP (1978) Attitudinal qualities relating to the strength of the attitude-behavior relationship, Journal of Experimental Social Psychology 14: 398-408.

15. Fazio RH, Zanna MP, Cooper J (1978) Direct experience and attitude-behavior consistency: An information processing analysis. Personality and Social Psychology Bulletin 4: 48-51.

16. Regan DT, Fazio R (1977) On the consistency between attitudes and behavior: Look to the method of attitude formation,Journal of Experimental Social Psychology 13: 28-45.

17. Schmitt B (1999) Experiential marketing,Journal of Marketing Management 15: 53-67.

18. Davidson AR, Jaccard JJ (1979) Variables that moderate the attitude-behavior relation: results of a longitudinal survey, Journal of Personality and Social Psychology 37: 1364-1376.

19. Norman DA, Bobrow DG (1975) On data-limited and resource-limited Processes. Cognitive Psychology 7: 44-64.

20. Warner LG, DeFleur ML (1969) Attitude as an interactional concept: Social constraint and social distance as intervening variables between attitudes and action. American Sociological Review34: 153-169.

21. Snyder M, Swann WB (1976) When actions reflect attitudes: The politics of impression management. Journal of Personality and Social Psychology 34: 1034-1042.

22. Fazio RH, Williams CJ (1986) Attitude Accessibility as a Moderator of the Attitude-Perception and Attitude- Behavior Relations: An Investigation of the 1984 Presidential Election. J PersSocPsychol 51: 505-514.

23. Lin CP, LL Tang, YB Chiu, Hsiao CY (2005) Testing a joint moderator of ego strength and ethical climate: a study of the process of peer reporting intentions in IT ethics, Asian Pacific Management Review 10: 145-153.

24. Chen CT, Chen CF (2011) The influence of internship experiences on the behavioral intentions of college students in Taiwan, The Asia-Pacific Education Researcher 20: 73-92.

25. Anderson JC,Gerbing DW (1988) Structural Equation Modeling in Practice: A Review and Recommended Two - Step Approach. Psychological Bulletin 103: 411-423.

26. Nunnally JC (1978) Psychometric theory. (2ndedn), McGraw-Hill, New York.

27. Bentler PM, C-P Chou (1987)Practical issues in structural modeling. Sociological Methods & Research 16: 78-117.

28. Hair JF Jr, Anderson RE, Tatham RL, Black WC (1998) Multivariate Data Analysis. (5thedn), Prentice Hall, Upper Saddle River, NJ.

29. Joreskog KG, SorbomD (1982) Recent Developments in Structural Equation Modeling. Journal of Marketing Research 19: 404-416.

30. Hu L, Bentler PM (1999) Cutoff criteria for fit indexes in covariance structure analysis: Conventional criteria versus new alternatives. Structural Equation Modeling: A Multidisciplinary Journal 6:1-55.

31. Carmines EG, McIver JP (1981) Analyzing Models with Unobserved Variables: Analysis of Covariance Structures. In BohrnstedtGW, Borgatta EF (eds) Social Measurement: Current Issues.Sage Publications, Beverly Hills, CA.

32. Tucker LR, Lewis C (1973) A reliability coefficient for maximum likelihood factor analysis. Psychometrika 38: 1-10.

5

HOTEL REVENUE MANAGEMENT STRATEGIES DURING FALL FOLIAGE TRAVEL SEASON

Stephen C Morse1* and Eva M Smith2

1Hospitality and Tourism Management, College of Business, Western Carolina University, Cullowhee, NC, USA

2Business Organization and Leadership, Brevard College, Brevard, NC, USA

ABSTRACT

Tourism during the fall seasons is mostly recognized during the months when changes in leaf color offer bright colors of many trees and foliage marking the end of the growing season in some regions of U.S. Hotel pricing analysts have increasingly used revenue management (RM) techniques and pricing strategies in hotels to assist in maximizing revenues during these seasons. This study examines how fall foliage tourism demand impacts hotel pricing strategies and hotel revenue management practices in areas experiencing seasonal increases in overnight travelers. The study area is the region around the Great Smoky Mountains National Park bordering both Tennessee and North Carolina. The area includes a portion of the Appalachian Mountains and is the most visited National Park in the United States. Most hotel revenue management strategies focus on one hotel property and not hotel pricing strategies in a region or area like this study.

KEYWORDS

Revenue management; Tourism; Management practices

1. INTRODUCTION

In some regions of the U.S., tourism generated by changing colors of the fall foliage season has grown to be one of the most important parts of the travel season. Tourism during the fall seasons is mostly recognized during the months when changes in leaf color offer bright colors of many trees and foliage marking the end of the growing season. Depending on the area of the U.S., this change in foliage color in the United States occurs mainly during the months of September, October and November each year. Travel and tourism increases each year in key areas of the U.S where fall color changes are most dramatic. This seasonal increase in travel and tourism in these key fall foliage areas increases overnight travelers and increases in the demand for hotel rooms in these areas.

This study examines how fall foliage tourism demand impacts hotel pricing strategies and hotel revenue management practices in areas experiencing seasonal increases in overnight travelers. The study area is the region around the Great Smoky Mountains National Park bordering both Tennessee and North Carolina. The area includes a portion of the Appalachian Mountains and is the most visited National Park in the United States with 9.68 million visitors in 2013 [1]. Most hotel revenue management strategies focus on one hotel property and not hotel pricing strategies in a region or area like this study.

1.1 Objectives of the Study

The objective of this study is to examine hotel room rates in a region during the fall foliage travel season to determine the degree that revenue management pricing strategies are used by hotels in the region during this period.

2. BACKGROUND AND LITERATURE REVIEW

Hotel pricing analysts have increasingly used revenue management (RM) techniques and pricing strategies in hotels to assist in maximizing revenues [2,3]. RM (and earlier called yield management) has been defined by Kimes as the process of allocating the right type of capacity to the right kind of customer at the right price so as to maximize revenue or yield [4].

Studies on the fall travel market and corresponding hotel pricing strategies are limited. Spotts and Mahoney [5] used an intercept survey of tourists in Michigan's Upper Peninsula to determine if there are differences in motivations and recreation activities of summer vs. fall tourists in the area. Spencer and Holecek [6] used a telephone survey of households in the Great Lakes region of Michigan to examine if fall tourists differ from summer tourists and whether fall color viewing was an important motivator of travel as assumed and if fall tourists travel for recreational activities. This study found that fall tourists were older, less active in recreation, and less likely to travel with children. Neither study [5,6] examined the revenue management and hotel pricing strategies of hotels during the fall tourist season and associated fall demand for tourism.

2.1 Hotel Price Elasticity of Demand

The concept of price elasticity of demand explains how sensitive consumers are to changes in prices. Price elasticity of demand has been defined as "a measure of the percentage change in quantity demanded resulting from a one percent change in price along a given demand curve" (higher or lower) [7]. When examining the market for hotel guests, they are usually grouped into two submarkets: leisure travelers and business travelers [8]. In order to accommodate the demand of these different market segments from business and leisure travelers, a hotel property must have a mix of different types of hotel rooms [2]. Business travelers in general tend to not be sensitive to hotel price rate fluctuations, leading to more inelastic demand for hotel rooms. However, leisure travelers tend to be more sensitive to hotel price rate fluctuations, leading leisure travelers to be more elastic demand for hotel rooms [8].

Economic theory suggests that there are many factors that determine if a good or service has an elastic or inelastic price elasticity of demand.

Hyman suggests there are three main factors that determine price elasticity of demand: a) the availability of substitutes, b) the proportion of the consumer's budget it takes to purchase the product, and c) the time period for consumers to adjust to price changes [7]. Based on these three factors, the demand for most good and services can be classified as being price elastic (sensitive demand) or price inelastic (not sensitive demand). Goods with many substitutes, goods that take up a large portion of the budget, and goods that have a longer time period of adjustment, tend to have more elastic demand; while goods with few substitutes, goods taking up a smaller portion of the budget, and goods that have a shorter time period of adjustment, tend to have more inelastic demand [9].

In many areas, fall foliage colors have a relatively short time horizon, sometimes lasting only a few weeks. Experienced leisure travelers seeking fall foliage color changes know there are few substitutes available for fall foliage color changes. Thus, for the few weeks of peak fall color for leisure travelers, the price elasticity of demand for hotel rooms during this limited period could be considered "price inelastic" (or not very sensitive to changes in prices). In this example, instead of business travelers having more inelastic demand, fall foliage travel during this limited season with few substitutes could see leisure travelers having more inelastic demand. To examine how hotel revenue management and pricing strategies react to this limited period of fall color, aggregate hotel rates are examined in the Great Smoky Mountains National Park area during the peak fall periods and compared to other parts of the year.

3. METHODOLOGY AND DATA SOURCE

The area of analysis is the 21 western counties in North Carolina that border the Great Smoky Mountains National Park and are no more than 50 miles from the Great Smoky Mountains National Park as seen in Figure 1. Data from Smith Travel Research (STR) were obtained for aggregated hotels in the area that were divided into five regions [10]. The data includes monthly room demand and Average Daily Rate ($ ADR) for 2012. Hotel performance data were aggregated for each region using the following county groups.

Figure 1. Regions of North Carolina County Groups examined for fall foliage travel and hotel pricing strategies.

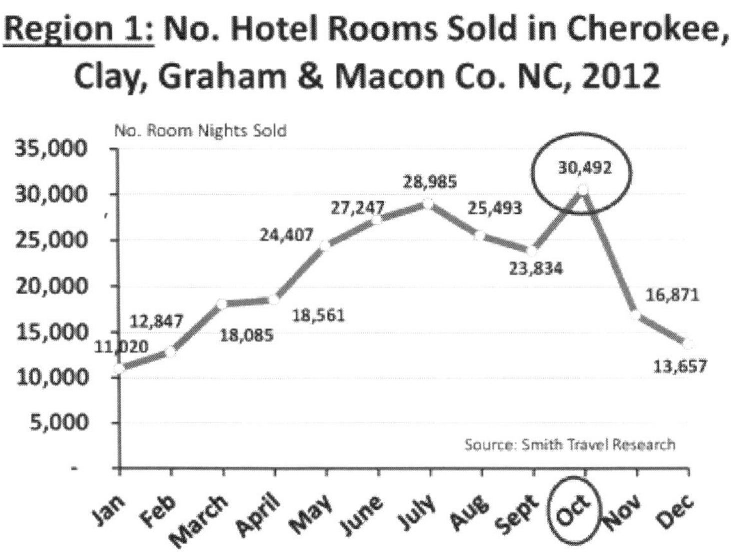

Figure 2. Region 1 number of hotel rooms sold monthly 2012. Source: Smith Travel Research

Region 1 Counties: Cherokee, Clay, Graham, Macon

Region 2 Counties: Haywood, Jackson, Transylvania, Swain

Region 3 Counties: Avery, Ashe, Wilkes, Alleghany, Caldwell, Watauga

Region 4 Counties: Yancey, Mitchell, Madison, McDowell, Burke

Region 5 Counties: Buncombe, Henderson

The methodology used will examine each region for number of rooms sold each month for 2012. It was determined that using the number of rooms sold each month is the best measure of tourist demand instead of hotel occupancy rates that can change if the supply of rooms change in the area. By using the number of rooms sold, the importance of the fall foliage travel season can be compared on a similar basis region to region.

Also, monthly hotel average daily rate (ADR) was used to show revenue management strategies of hotels in each region and demand for hotels rooms change during the fall foliage season.

Economic theory and revenue management strategies that maximize hotel revenues suggest that hotel room rates change directly with demand. That is when the demand for hotel rooms are highest, room rates should be highest and when the demand for hotel rooms are lowest, hotel room rates should be lower. Monthly data from Smith Travel Research indicating both demand for hotel rooms (i.e. number of rooms sold), and average daily rates charges by area hotels each month allows us to examine if seasonal hotel revenue management practices and strategies are being executed.

4. RESULTS

The importance of the fall foliage season on tourism demand during 2012 is shown for each region in Figures 2-6. The results show that in each region, October is the highest month for number of hotel rooms sold. In this area, October is the peak color month for fall foliage and thus the most popular month of travel out of all 12 months of the year.

Since October was the highest month of hotel room sales, hotel revenue management price strategies were reflected when hotel average daily rates are analyzed monthly for 2012. Figures 7-11 show the ADR for each region and indicate that October has the highest ADR's out of each month of the year.

When demand and ADR data are combined, it show that in October, the hotel sector uses revenue management strategies to maximize revenues during the fall foliage travel season in this area of the U.S.

Figure 3. Region 2 number of hotel rooms sold monthly 2012. Source: Smith Travel Research.

5. DISCUSSION

Economic theory and revenue management maximization strategies conclude that hotel room rates should be increased during periods of consumer inelastic demand. In these data, fall foliage tourist travel during limited peak viewing times suggest that tourists have few substitutes during the fall foliage color period and thus have more inelastic demand (less sensitive to prices) for travel and hotel rooms and are willing to pay higher room rates during this period.

Figure 4 Region 3 number of hotel rooms sold monthly 2012. Source: Smith Travel Research.

It should be pointed out that this special case of fall foliage travel goes against conventional wisdom and experiences of most hotel revenue pricing strategies where in general, most business travelers are considered to have more inelastic demand (less sensitive to higher prices) for hotel rooms, while in general most leisure travelers have more elastic demand (more sensitive to higher prices) for hotel rooms. Thus is this study of fall foliage travelers where few substitutes exist because of the short time frame and window of peak fall colors, most leisure travelers have more inelastic demand for hotel rooms and hoteliers use appropriate revenue management strategies to increase hotel room rates during the period of increased leisure travelers having inelastic demand during the short fall foliage travel season.

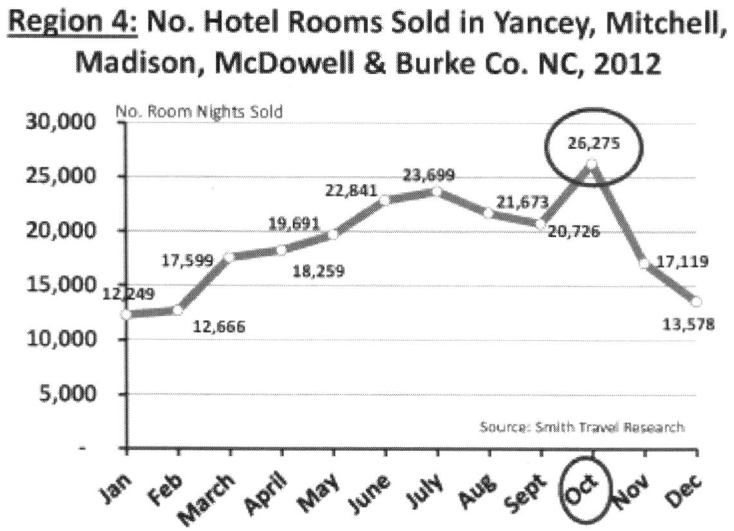

Figure 5. Region 4 number of hotel rooms sold monthly 2012. Source: Smith Travel Research.

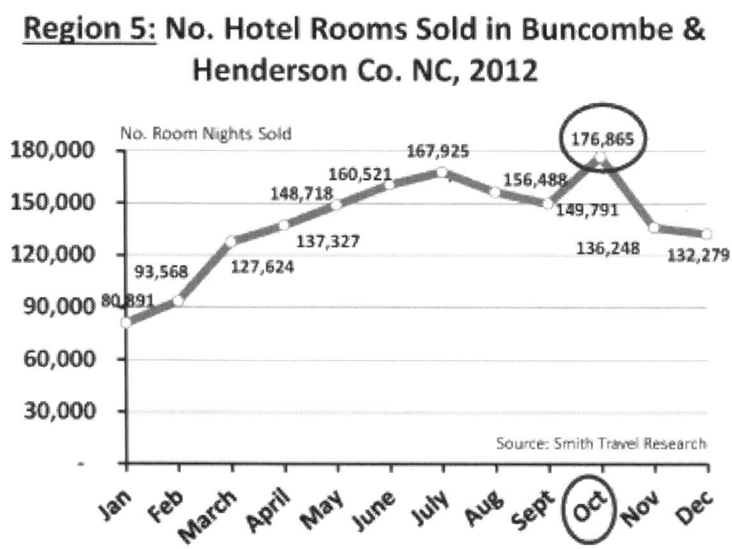

Figure 6. Region 5 number of hotel rooms sold monthly 2012. Source: Smith Travel Research.

Figure 7. Region 1 hotel average daily rates, monthly 2012. Source: Smith Travel Research.

Figure 8. Average daily rates, monthly 2012. Source: Smith Travel Research.

Figure 9. Region 3 hotel average daily rates, monthly 2012. Source: Smith Travel Research.

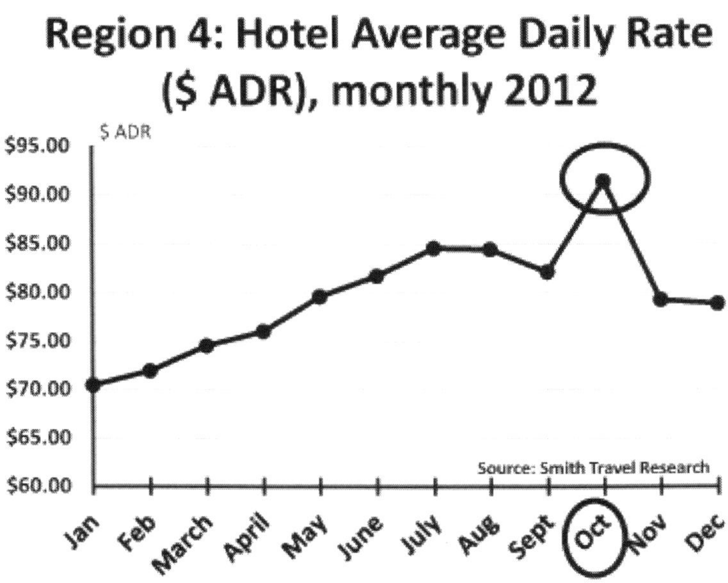

Figure 10. Region 4 hotel average daily rates, monthly 2012. Source: Smith Travel Research.

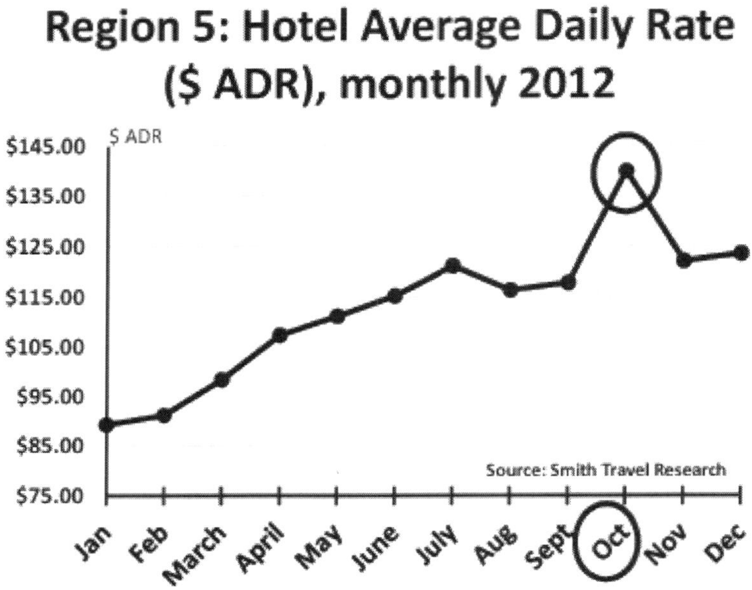

Figure 11. Region 5 hotel average daily rates, monthly 2012. Source: Smith Travel Research.

REFERENCES

1. U.S. Department of Interior (2013) National Park Visitation, Great Smoky Mountains National Park.

2. Kime SE (2003) Revenue management: A Retrospective. Cornell Hotel and Restaurant Administration Quarterly 44: 131-138.

3. Orkin EB (1988) Boosting your bottom line with yield management. The Cornell Hotel and Restaurant Administration Quarterly 28: 52-56.

4. Kimes SE (1989) The basics of yield management. Cornell Hotel and Restaurant Administration Quarterly 30: 14-19.

5. Spotts DM, Mahoney EM (1993) Understanding the fall tourism market. Journal of Travel Research 32: 3-15.

6. Spencer DM, Holecek DF (2007) Basic characteristics of the fall tourism market. Tourism Management 28: 491-504.

7. Hyman DN (1989) Modern microeconomics: Analysis and Application.(2ndedn), Irwin Professional Publishing, Homewood.

8. Relihan WJ (1989) The yield management approach to hotel room pricing. Cornell Hotel and Restaurant Administration Quarterly 30: 40-45.

9. Leftwich RH, Eckert RD (1982) The Price System and Resource Allocation (8thedn), The Dryden Press. Chicago.

10. Smith Travel Research (2014) Hotel data on selected county areas in North Carolina 2008 – 2013, Hendersonville, TN.

6

THE ROLE OF BRANDING, PROMOTION AND SUB-CULTURE IN THE CONSUMPTION OF BREAKFAST CEREALS IN THAILAND

Christina Koutra*, Monlapak Thespol and Isaac K Ngugi

The Business School, Executive Business Centre (5th floor), Bournemouth University, Bournemouth, UK

ABSTRACT

Nowadays, firms try to comprehend crucial factors impacting on consumer behaviour to sustain themselves in the market. Although these factors may vary and depend on many external and internal environmental influences this study examines the role of branding, promotion and sub-culture in influencing the consumption of breakfast cereals in Thailand through a qualitative, interpretive approach using as a case study Kellogg's a well-known international cereal breakfast company. Grounded theory used to analyse findings suggests: branding influences consumer's purchase decision making in Thailand while age is a contributing factor to consumers' choices and especially with reference to breakfast cereals. Hence it is important that firms keep adjusting their marketing strategy in line with the changing consumption patterns if they want to keep themselves ahead of the game.

KEYWORDS

Branding; Sub-culture; Promotion; Consumption; Breakfast-cereals; Thailand

1. INTRODUCTION

Several factors influence consumer behaviour [1-3] and they tend to vary from industry to industry. This is mainly due to internal and external environmental changes that globalisation has brought about worldwide. Accordingly, it is important that international organisations comprehend consumers' preferences in different parts of the globe and adopt strategies that boost demand for their products and/or services [4]. With respect to food, the differences in consumption across countries may arise from divergences in consumer characteristics such as nationality, religion, education, sex, attitudes and behaviour [5]. Thus, people who live in different countries would likely have different patterns of consumption [6]. Therefore, it would be difficult for international companies to realise the needs of all consumers worldwide with the same products/services and using the same marketing strategies. Firms may have to adjust their products to comply with certain consumption patterns in local markets. Brynes suggests that organisations and marketers in particular should focus on comprehending the local and cultural component of the countries they target. Das states that globalisation has greatly influenced consumption patterns, whilst Ravallion stresses the role of the evolution of living standards within a country, as indicated by its Gross Domestic Product (GDP). For example in Asian countries, the annual GDP increased from $765 to $4,359 between 1950 and 2000 [7,8]. In Thailand specifically, since the 1960s the GDP has steadily increased due to economic growth. From 2007 to 2008 alone the GDP saw an eight percent increase. Further, Asian countries also score high on the uncertainty avoidance index, an indication that Asians prefer to choose products they can trust and are certain of their quality [9-11].

This study focuses on the influence of branding, promotion, and sub-culture on food, with emphasis on branded breakfast cereal in Thailand, and on consumers' decision making processes, in order to assess patterns of behaviour. Accordingly, a) types of consumer behaviour are explored with a view to identify their origins; b) the possible links between branding and consumer behaviour are examined; c) the difference between consumption patterns of the old and new generations are investigated to identify possible influences of sub-culture on consumption of breakfast cereals; d) and the most effective marketing communication tool in influencing consumption of breakfast cereals in Thailand is considered to establish reasons for its success.

To enable the investigation of the role of branding, sub-culture and promotion in the consumption of breakfast cereals, this research focuses on the case of Kellogg's breakfast cereals. Kellogg's, which is regarded as the leading breakfast cereals in the world, uses four main types of promotional tool, namely: public relations, advertising, sales promotion and direct marketing [12-17]. However, the extent to which branding influences purchase, which of these tools best promotes the breakfast cereal in Thailand, and the effect of sub-culture have not, so far, been established; an issue which is addressed in this study.

The study adopted a qualitative interpretive approach to research within a case study strategy in order to investigate food consumer behaviour with regard to breakfast cereal and most specifically the Kellogg's brand. The findings indicated that consumer patterns of behaviour changed due to a shift in the economy from agricultural to industrial, a competitive work environment, traffic congestion, a concern for a healthy lifestyle, preference for a branded product, advertising and sub-cultural influences.

Primary market data may involve qualitative research or quantitative research. Both types of data are valuable in understanding what consumers want or need. Qualitative data is concerned more about opinions, feelings and attitudes. Quantitative data is in numerical form and is usually gathered from a large sample of respondents.

Qualitative research establishes a conversation with consumers. It prompts consumer reaction to, for example, a new product idea and helps researchers understand what they think of it, how it makes them feel, why they find it interesting or not.

Quantitative research may use questionnaires administered to large numbers of respondents. This allows statistical analysis, such as the calculation of a mean score or percentages. It aims to give a representative picture of what consumers think of a new product idea or a new (real) food.

As this research focuses on development of Crunchy Nut Bites, a more recent extension to the Crunchy nuts brand. It was important for

Kellogg's to conduct both Qualitative and Quantitative research because by conducting qualitative research it was able to know the demands of the consumer in terms of potential flavours and textures.

The results of the qualitative research allowed the Kellogg's to discard some ideas. Other ideas were appealing to consumers but needed refining and further development.

Quantitative research was also important. As Kellogg's came out with a number of products to produce, these products were shown to a number of large group of representative consumers in a quantitative survey. They were asked to rate those ideas against a number of scales, so Kellogg's could identify which product ideas consumers liked best or disliked.

2. CONSIDERING BRANDING, PROMOTION, AND SUB-CULTURE AND THEIR LINKS WITH CONSUMER BEHAVIOUR

According to Doyle, a '[...] successful brand is a name, symbol, design, or some combination, which identifies the "product" of a particular organisation as having a sustainable differential advantage'. 'Successful brands are normally brand leaders in their segments. Brands work by facilitating and effecting customer's choice process' Doyle. Consumers will usually buy brands that they have used before and will continue to use them as long as their satisfaction level is high. Nevertheless, social, cultural and personality factors may influence their choice for a specific brand. Additionally, with regard to low involvement products, consumers usually switch easily to the substitute brands, if preferred brands are not available (ibid).

In recent years, with respect to breakfast cereals, there has been an increase in "own-label" (private label) cereal products alongside traditional brands, which makes it difficult for consumers to distinguish between them [18,19]. From this point of view, the quality increase of the private label brands could be a major factor which reduces the differentiation between brands. Therefore, an opportunity arises for cereal companies to adjust their marketing strategy appropriately in order to differentiate themselves among competitors' brands. One way to do that is by considering sub-cultures.

One way of defining sub-cultures is along generational lines or rather segmentation by age. People in the same age-group tend to have similar life-experiences, needs and also consumption patterns. Marketers tend to classify consumers into four age-groups: Generation Y/millenniums, Generation X, Boomers and Seniors [20,21]. In particular, Generation Y is comprised of people who were born between the years 1977-1994. This generation likes to watch television programmes, listen to music, play sports and use other communication tools such as the internet [20]. They are also the main group that use communication media, such as Facebook, Twitter and YouTube [21-24]. This generation, because of their constant exposure to new products and services, consists of the most popular target-consumers for marketing communications.

Further, Generation Y tends to spend large amounts of money on items such as computers, cars and clothes and their purchasing behaviour is easily influenced by famous people such as sports celebrities and/or actors/actresses [25]. With reference to Asia, Davis points out that Generation Y, having grown up with the digital period, is presently using various channels of communication such as internet communities and advanced mobile phones and these can have a strong impact on their behaviour and perceptions [26]. Bareham, Peter and Olson suggest that the culture in which people were brought up plays a critical part in consumption patterns and in turn has implications for the communication channels that companies use in order to create an appropriate impact [27,28].

In recent years, there has been an increasing interest in marketing communications in global business organisations. This is because by using the appropriate tools via the right channels, they can assist toward communicating a product directly to the target consumers. According to Baines et al. and Constantin, the wide range of tools and media which comprise the marketing communications mix, and include advertising, sales promotion, public relations, direct marketing and personal selling, are used by many organisations for attracting their target customers [29,30]. However, at present, innovative technologies, in the form of the internet, websites and mobile devices, alongside the traditional communication tools, have increased in their efficiency and are simultaneously providing product information worldwide [31]. Moreover, technologies should not only be integrated with the overall information systems strategy but should also be combined with the whole marketing strategies of their organisations [32].

In the case of marketing communications, the basic elements which are always used as the promotional mix for promoting products and services are the 4Ps of marketing mix: product, price, place and promotion [33]. Furthermore, there are several tools, such as advertising, sales promotion, public relations and direct marketing, which marketers can choose from. They can choose the most appropriate element or combine various tools together for more effective outcomes. Usunier and Lee point out advertising as the key tool of marketing communication which is usually used to release the message to the target audiences [34]. However, marketers should not only place an emphasis on the marketing communication tools but they should also focus on the target consumers and ensure that they receive and recognise their messages. As Ballantyne points out, the communication is not only concerned with the media communicating to its audience but also with marketers communicating with their audiences [35].

Schiffmann and Kanuk [3] suggest that the principal components of communication are sender, receiver, medium, message and feedback. They explain that all these components play an important role in marketing communications. For example, they argue that the medium, which is the channel of communication, is important as the marketers have to select the most effective channel through which to reach their target consumers; whereas feedback is important because the message, which is sent to receivers, can be verified by consumer feedback which confirms (or otherwise) that the message actually conveyed is correct.

In relation to cultural differences, it is important for marketers to learn and understand local behaviours that impact on consumer demand [36]. An organisation's competitive advantages will be enhanced by the provision of appropriate products and services that are adapted to local usage conditions. However, although advertising is the most effective tool for international marketing communications, it has, on occasion, been associated with advertising campaign miscommunication, against a backdrop of cultural, language and behavioural differences between countries. Indeed, Blythe points out that there are few ads which can be launched in any country, and where there is no need to change anything except the language. For these reasons, it can be assumed that the best way to reach target consumers in foreign countries is to learn and to understand their cultures, languages and behaviours in order to effectively and efficiently reach them and to increase profitability in global markets.

Above all, Mooij points out that marketers should be concerned about the advertising styles to use in different countries to communicate their products because the appropriate method of advertising will usually vary across cultures. The study suggests various components of advertising styles and these can be implemented differently in different cultures. They include: appeal, communication style (e.g. explicit, implicit, direct, indirect), basic advertising form (e.g. testimonial, drama, entertainment) and execution (e.g. how people are dressed).

Ghauri and Cateora suggest that in order to reach the target markets in different cultures, it would be better for international firms to hire local advertising agencies or multinational advertising agencies which have local branches in different countries and have adequate experience in the target countries [37]. As the local advertising agencies should have more knowledge about the local people in each country than the organisation, the agencies would more likely create an effective campaign which can reach the target consumers and increase the profitability of the companies by using the right channel and content in communications [38,39].

3. THAILAND AND CONSUMER BEHAVIOUR

The study area for this research is Thailand. Thai people are traditionally known to prepare and cook all meals of the day for themselves at home. They eat rice at almost every meal and their breakfast is customarily based on soy milk, deep-fried dough sticks and congee (boiled rice) [40]. However, in recent years, their consumption pattern seems to have changed. Hence, increasingly people have started to eat out and/or to buy ready-made food rather than preparing their meals at home (National Research Council of Thailand). This change in behaviour is partly attributable to the economic downturn and globalisation, and consequently busy lifestyles [41]. For example, due to the economic downturn, people were forced to work longer hours – waking earlier and returning home later. Hence, they need to take a quick but nutritious breakfast [42]. At the same time, people are increasingly interested in healthy lifestyles and healthy food with increasing demand for fruit and vegetable juices (ibid). All these signify changes in consumption behaviours. In recent decades Thailand has also become more industrialised as the country increasingly changes its economic structure from an agricultural to an industrial one. As a result, people have to work harder than in the past, which can lead to spending less time on domestic chores such as cooking, but also to less time with their families [43].

4. METHODOLOGY AND RESEARCH DESIGN

A qualitative inductive approach to research within a case study strategy was adopted since the aim of the study was to comprehend the way actors understand the social world and the meaning behind their actions; in this instance, to comprehend the way the consumers' purchasing behaviour is influenced by the role of branding, marketing communications and sub-culture in the consumption of breakfast cereals with special emphasis on Kellogg's in Thailand. The participants consisted of Thai people who consume Kellogg's breakfast cereal and were born between 1977 and 1994. They are members of generation Y and are mainly characterised by their intense exposure to communication, media and digital technology [44,45]. In total 15 people were interviewed using an open-ended semi-structured set of interview questions which allowed the researcher to probe with more questions hence eliciting more information. The interviewing processes ceased when saturation was achieved. Given that the sample had to comply with the specific characteristics of age and consumption of the Kellogg's cereal breakfast brand, a non-probability, purposive, technique was used.

The interviewees were identified both in Thailand and the United Kingdom and were interviewed either face-to-face (for Thai respondents who resided in the UK) or via telephone and Skype (for respondents who resided in Thailand). All interviews were conducted in the Thai language because it gave more room for discussion as the respondents felt more comfortable speaking in their mother tongue. The interviews were recorded using a voice recorder. The voice recorder assisted the researcher to refer back to the conversations when necessary. Then the conversations were translated from Thai to the English language and then analysed.

Grounded theory was used to analyse the data since, as Saunders states it can help the researcher to discover a wide range of business and management issues such as consumer behaviour [46]. Further, since the aim of the research was to come up with certain prepositions, the use of grounded theory seemed an appropriate analytical process due to its rigour and systematic nature of coding: open, axial and selective.

Open coding involves 'breaking down, examining, comparing, conceptualizing, and categorizing data often, in terms of properties and dimensions [47]. Consequently, the data were firstly codified, having the aim and objectives in mind and asking open questions, such as why?, how? and what? Then the codes were grouped into concepts according to meaning and relevance to the literature. The following concepts were identified and classified: 'marketing communication', 'promotion', 'tools', 'change on consumer behaviour', 'change', 'culture', 'new generation', 'difference', 'purchasing decision-making', 'brand', 'food consumption', 'breakfast cereal', 'Kellogg's', 'traditional', 'globalisation', 'patterns', 'international', 'profitability' and 'cereals'.

Then the concepts were grouped into categories following repetition patterns and relationships among concepts as part of the axial process .The following categories were formulated: 'branding', 'consumption patterns', 'marketing communications tools (advertising, sales promotion)'. Then the categories were extended into prepositions considering themes to be discussed as well as relations between those themes (Table 1).

The final step in grounded theory, selective coding was not used since the aim was not to induce a theory per set but to come up with certain prepositions that best discusses the aim and objectives of the research.

Table 1. Axial coding

Category	Prepositions
Consumption Pattern	An economic shift from agriculture to industry brought about changes in food consumption patterns in Thailand. Branding and promotional channels contributed to a change in food consumer behaviour
Branding Marketing communication tools (advertising, sales promotion)	Comprehension of sub-cultures within a culture is essential for the effective use of marketing communication tools to sell breakfast cereals.

5. ANALYSIS AND FINDINGS

5.1 An economic shift from agriculture to industry brought about changes in food consumption patterns in Thailand.

Consumer behaviour in Thailand has seen '[...] many changes over the last few decades. Especially due to the economy's shift from agricultural to an industrial one' said an interviewee. Kosulwat points out the change of economic structure from agricultural to industrial as the major reason for economic growth in Thailand since the 1960s, and development, it could be argued, depicted at a societal level in terms of knowledge, education, and information dissemination [48]. In view of that, the change in food consumption pattern, such as the change in expenditure on food products from home prepared food to ready-to-eat food, is due to a number of factors including globalisation, enhanced living standards, economic growth/development, and a change in lifestyle and behaviour of the people of the country (ibid). This research reinforces this view and it also informs knowledge by pointing out that a competitive work environment, traffic, and a concern for a healthy lifestyle are also contributing factors to changing patterns of consumer behaviour in Thailand.

Accordingly, people's habits have changed in an effort to adjust to a more competitive work environment associated with longer working hours. One interviewee said '[...] due to high competition at the work place, I have to work harder to get promotion than in the past. Therefore, I have to spend most of my time at work and do not have enough time to cook anymore [...]'. While another one added 'I have to work more than eight hours a day for six days a week, so I do not have enough time to cook, and I resort to buying ready-to-eat or frozen food [hence] most of my salary will be spent on food'. In relation to this, Shannon (2008) maintains that nowadays people tend to have less time to cook due to the time-pressure arising from long working hours. Meanwhile, Euromonitor International argue that, due to continued economic growth and the change of lifestyle in Thailand, people tend to spend a larger proportion of their incomes on food consumption [49].

Another contributing factor to changes in food consumer behaviour is time spent commuting. '[D]ue to the traffic congestion every morning, I do not have enough time to cook, therefore, I have to eat within five or ten minutes and then go to work [...]; also, I have to spend time for transportation, more than two hours every day. [...]. I think we have very bad transportation in Bangkok. Therefore, I have to stay in my car more than two hours every day, which means I have less time for anything else [...] so cereal breakfast is a fast solution for that' said some interviewees.

A concern for healthy well-being is also a contributory factor. One interviewee said 'Even if I do not have enough time to cook in the morning, I still want to eat food that can give me the nutrients and vitamins [...] some types of breakfast cereal are reputed for that however I prefer the branded ones and especially Kellogg's since I have more trust in the manufacturers". Whilst another one added [...] I would prefer to choose breakfast cereals which contain appropriate types of vitamins and nutrients [...]. I always read nutrition details before purchase because I want to know what kind of nutrients I can gain from the products [...]. I choose my favourite breakfast cereal based on the ingredients, i.e. whole-grain [...] I trust Kellogg's on this'. Thompson and Moughan, Rahavi and Kapsak note that increasingly people are placing an emphasis on health and wellness rather than on other things because this provides several beneficial effects such as reduction in diseases and illnesses. Likewise, there is an increasing consumer awareness about human health and environmental issues and concern for food safety, quality and security [50-52].

5.2 Branding and promotional channels as contributing factors to a change in food consumer behaviour

Branding also influences food consumer behaviour. Thai people prefer to purchase well-known brands, and this is usually attributed to their reputation for quality and value for money. An interviewee said 'With regards to the purchase of basic products that are available everywhere in the market such as chilli, basil leaf and egg, I would choose the local ones due to the lower price. However, for particular products such as chocolate, breakfast cereal and rice, I would prefer the famous brands because I need more quality'. Accordingly, there is a link between consumers' behaviour and branding. This complies with Hofstede who argues that the purchasing of famous brands is a characteristic of collectivist society (and Thailand is at a high level on the collectivistic index), unlike individualistic cultures where people are more likely to purchase unknown brands [53]. Furthermore, Thai people do not feel that they can receive the same quality from unknown brands as their trust is mainly based on the reputation of the wellknown brands [54].

Further another interviewee added 'In order to compare the same product between the famous and local brands, I would consider the price of the famous brands first and if it is worth it, I would be willing to pay for the better things [...] my decision is mainly based on its brand [...] I trust well established brands to give the quality I am after and I am willing to pay for this' said some interviewees.

These findings contradicts Rickwood and White study that argues that financial status determines purchasing behaviour and that the lower priced products may attract more consumer attention, and Bareham study which argues that household income and price are issues that affect consumer behaviour and the demand of consumption as well.

However Kellogs in order to address also low income households and consequently increase even further sales uses promotional channels such as sales promotions, and advertising. One interviewee said 'I have known Kellogg's from the promotions in the supermarket such as buyone- get-one-free and those that give out free toys or bowls along with a purchase of a cereal box. While another one added '[...] due to the high price of the breakfast cereal in Thailand, I usually buy many boxes at the same time since they are cheaper this way [...] Kellogg's does that occasionally [...] the price of Kellogg's cereal in Thailand is quite high if we compare with the average incomes of Thai people. For example, the price of Kellogg's Frostie 300 grammes is about 100 baht (£2). However if the company has the effective promotion, such as buy one get one free or half price, it would increase the purchase volume from the consumers in Thailand'.

In the past years, Kellogg's Company has used sales promotion to boost up sales and brand value such as price promotions by discount price of products per unit within specific period, premium promotions such as provide free gifts within cereal boxes and joint promotions such as get free a pint of milk and a box of cereal when consumers can completely collect special codes.

Kellogs is also using advertising to increase its visibility 'I have known Kellogg's breakfast cereal for ten years as I watched the advertising on TV during the cartoon programme. Afterwards I asked my parents to buy it because I was impressed with the presenter of the product, which is the Orange Tiger'. The influence on consumer decision-making is the key target of advertising, and one of the most important types of communications [55,56] point out that advertising is the most powerful tool of global marketing at present; it is still a difficult task to launch the same advertising all over the world in order to reach target consumers worldwide due to the cultural differences and behaviour across countries.

Direct marketing and public relations even though used from the company do not seem to have an impact on consumers decision making since none of them seem to have. Beyaztas argues that the company used direct marketing since 1998, and as a result of which over 1.5

million customers were alerted through direct mail. One interview said 'I am not so sure about direct advertising usually it is social media that we get our information from'. This can be tribute to the fact that Beyaztas study has taken place when social media were not used as a means for advertising as well as the fact that the sample of people were coming from a one city only even though the capital of Thailand. With regards to public relations as a promotional tool according to Kellogg tries to increase brand awareness in Thailand through a cooperation with the Hard Rock Hotel in Pattaya; the hotel is promoted on Kellogg's package.

From the arguments above, it is possible to say that the communications' tools, such as advertising, promotions and public relations, can be one of the influential factors which can affect on consumers' decision-making. However, as it has been mentioned earlier about the consumer diversity in the different countries, it is the major issue that should be concerned as the consumer decision-making would be affected from the communications' tools in the different ways in different countries. As a result of the data analysis in Thailand, it can be assumed that the promotional tool can affect on consumer decisionmaking of the Kellogg's company in Thailand.

5.3 Comprehension of sub-cultures within a culture is essential for the effective use of marketing communication tools to sell breakfast cereals

There is a clear distinction between the breakfast preferences of the younger and older generations and this reflects consumption differences across the sub-cultures. One interviewee said, '[...] my parents always eat a traditional meal, such as congee, every morning but I prefer breakfast cereal or toast'. According to De Mooij, in collective societies people are less prone to innovation. However when innovation is adopted by a group's culture or sub-culture, even though initially the rate of adoption might be slow, when it takes root the speed of diffusion is higher than in individualistic societies. Another interviewee commented '[...] no one in my family eats breakfast cereal for breakfast except me, my sister and brother'.

Accordingly, consumption preferences of the younger generation are different from those of the older generation. The older generation tended to consume traditional breakfast meals in contrast to subcultures such as generation Y which prefers breakfast cereals and readymade meals. Bareham and Peter and Olson suggest that the culture in which people were brought up plays a critical part in consumption patterns and in turn has implications for the communication channels that companies use in order to create an appropriate impact [57].

Further, another interviewee added, 'My father and mother and grandparents don't really trust new products, they prefer what they already know and are used to'. Rokkas argued that societies with high uncertainty avoidance do not trust new ideas and behaviours. Accordingly, marketers should take into account traditional lifestyles, consumers' age and local culture when marketing products. Since people are interested in health promotion, this should be around the health benefits that cereals might bring to both the younger and older generations. Pankhania argue that it is crucial to take into consideration not only cultural diversity within countries but also between them. This is because a group's identity might be stronger and more important for those who are a part of it, which might relate to the feeling of belonging.

6. CONCLUSION

The purpose of this study was to explore breakfast cereal consumption in Thailand while taking into consideration the role of branding, sub-culture and promotional communication tools. Kellogg's was used as a case study. Interviews were employed as the data collection method in order to obtain rich qualitative data from the respondents. The grounded theory analytical procedure was used in order to come up with prepositions which best discussed the data which were linked to the themes of consumer behaviour, branding, subculture and promotion.

Consequently, a shift from an agricultural based to an industrial based economy contributed to changes in food consumption; this is a shift from traditional to convenient breakfast meals with reference to breakfast cereal and especially among the younger generation. Branding is paramount when choosing a cereal product and Kellogg's is favoured by consumers since it is linked to healthy nutrients.

More specifically, the findings suggest that Thailand has witnessed many changes in the past few decades with regard to its economic structure, technology and society, among other things. These changes have impacted on consumer behaviour and societal consumption patterns. Long working hours and limited spare time have been identified as one of the most influential factors in the growing trend to substitute a fast, cereal based, breakfast for the traditional breakfast. Consumers prefer branded products with a preference for Kellogg's due to its well established name which to them translates into quality, trust and reliability. This emphasises the need for breakfast cereal companies to reinforce investment in branding whilst their marketing strategy should reflect societal changes in order to satisfy consumer needs, wants and desires, enabling them to take a share of the market.

Furthermore, due to innovations in media and technology, the most effective promotional tool is advertising. The results suggest that, because of advertising, consumer behaviour in relation to the food industry has seen a change towards healthier consumption as consumers' concerns about their health have increased in comparison to past decades. With regard to Kellogg's, respondents said that they prefer to purchase the brand as it is linked to a healthy lifestyle, nutrients and vitamins. Rahavi and Kapsak also argue that, due to the increasing importance attributed by consumers to health and functional foods it is important that food producers pay ample attention to the nutritional value of food products because these factors are crucial in influencing consumer decision-making on the food products that they purchase. Therefore, the marketing strategies of breakfast cereal companies which are linked to the above elements achieve high sales.

As far as the marketing communication strategy is concerned, the findings suggest that there is a strong relationship between change in consumption patterns, consumer behaviour, and sub-cultures which could have an influence on international marketing strategies for companies which aim to succeed in foreign markets. As people in different countries tend to exhibit different behaviours, attitudes and consumption patterns, marketers should learn to distinguish between cultures in order to launch the right products to people and receive positive feedback from their consumers.

7. LIMITATIONS AND OPPORTUNITIES FOR FURTHER RESEARCH

The research was undertaken in the capital, Bangok, and therefore the findings might be different in rural areas where the pace of life is slower and the effects of the new economic trends are not so visible. Therefore future research might see a comparison between urban and rural areas in order to identify ways that marketing can be used in order to penetrate untapped markets.

REFERENCES

1. Blackwell RD, Miniard PW, Engel JF(2006)Consumer BehaviorOhio Thomson Higher Education.
2. Blythe J (2008) Consumer Behaviour.London,Thomson Learning.
3. SchiffmanLG, Kanuk LL(2004)Consumer Behaviuor New Jersey Pearson Education Limited.
4. Mühlbacher H, Leihs H, Dahringer L (2006)International Marketing: A Global Perspective. London: Thomson Learning.
5. Tengpongsathon K, Nanthaicahi K, Mason R, Rimkeeree H(2006) Consumer Preference of Rice-based Products: Cross Cultural Study Between Thailand and Australia. Journal of Agricultural Technology 2: 137-153.
6. Hoyer WD,MacInnis DJ(2004)Consumer Behavior.Massachusetts: Houghton Mifflin Company.
7. Das DK (2009) Globalisation and an Emerging Global Middle Class. Economic Affairs 29: 89-92.
8. Ravallion M(2010) The Developing World's Bulging (but Vulnerable) Middle Class. World Development 38: 445-454.

9. Mooij MD (2003) Convergence and Divergence in Consumer Behaviour: Implications for Global Advertising. International Journal of Advertising 22: 183-202.

10. Hofstede G (2009)Geert HofstedeCultural Dimensions.

11. Euromonitor International(2009)Kellogg Co in Packaged Food - World.

12. Kellogg's(2009)Company Overview.

13. Magee K(2010)Kellogg's calls in Six PR to help launch youth cereal Krave.

14. Than Setthakij Newspaper (1999) Kellogg's increase promotion budget three folds. Thai News Service.

15. Kellogg's (2009)Press Office: Free Kellogg's Corn Flakes and a pint ofmilk.

16. Hot Wheels(2000)Kellogg's in the Hot Wheels Racing and Hot Wheels Pro Racing Series.

17. Beyaztas B (1998)Kellogg seeks loyalty with direct mail push.

18. Mintel (2010)Consumer Breakfast Eating Habits - UK. Available from: http://academic.mintel.com/sinatra/oxygen_academic//display/&id=479827/display/id=506878/display/id=508730&anchor=atom3#atom3 [Accessed 01 November 2010].

19. Hoyer WD,MacInnis DJ (2007)Consumer Behavior. 4th ed. Massachusetts: Houghton Mifflin Company.

20. Cheung LH, HarkerD(2008) The State of the Art of Advertising from the Consumers' Perspective: A Generational Approach.The Marketing Review 8: 125-146.

21. Paul P (2001) Getting inside Gen Y. American Demographics 23 : 42-49.

22. Winzar H,Swanepoel C,Lye A(2008)Virally Inspired: Gen Y Perceptions of Viral Stealth Marketing.

23. Ryan-Segger T (2007)The Why Generation. B&T Weekly,22-24.

24. Bush AJ, Martin CA, Bush VD (2004) Sports Celebrity Influence on the Behavioral Intentions of Generation Y. Journal of Advertising Research 44: 108-118.

25. Davis A (2007) Getting Down with Asia's Generation Y. Asia's Media & Marketing Newspaperpp:122.

26. Bareham J (1995)Consumer Behaviour in the Food Industry: A European Perspective. Oxford: Butterworth-Heinemann Limited.

27. Peter JP , Olson JC(2008)Consumer Behavior and Marketing Strategy. New York: McGraw-Hill Companies.

28. Baines P, Fill C, Page K(2008)Marketing. New York: Oxford University Press.

29. Constantin S(2009)The Role of Promotion in Milling and Bakery Products Sales. Theoretical & Applied Economics 16: 63-70.

30. Gaile-Sarkane E(2010) Diversity of Marketing – Trends and Developments. Economics & Management15: 478-482.

31. Fill C(2009)Marketing Communications: Interactivity, Communities and Content, Essex: Pearson Education Limited.

32. Hughes G, Fill C(2007) Redefining the Nature and Format of the Marketing Communications Mix. The Marketing Review 7: 45-57.

33. Usunier J, Lee JA(2009)Marketing Across Cultures. Essex: Pearson Education Limited.

34. Ballantyne D (2004) Dialogue and Its Role in the Development of Relationship Specific Knowledge. Journal of Business & Industrial Marketing 19: 114-123.

35. Dapkus G, Eimantaite G (2010) Adapting the Advertisement Considering International Markets Peculiarities. Economics & Management 15: 457-464.

36. Mooij MD (2010)Global Marketing and Advertising: Understanding Cultural Paradoxes. 3rd ed. London: SAGE Publications Limited.

37. Jeong J, Tharp M, ChoiH(2002) Exploring the Missing Point of View in International Advertising Management. Local Managers in Global Advertising Agencies. International Journal of Advertising 21: 293-321.

38. Faisal MN, Khan BM (2008) Selecting an Advertising Agency: A Multi-criteria Decision Making Approach. The Journal of Business Perspective 12: 13-22.

39. Loha-unchit K (2006)Congee (Rice Porridge) for Breakfast.

40. Euromonitor International (2010)Consumer Lifestyles – Thailand: Eating Preferences. Available from Euro monitor International.

41. Consumer Lifestyles – Thailand: Consumer Trends.

42. Ishizuka K, Hisajima S, Macer DRJ (1995) Changes in Tradition Society During Industrialization in Thailand. Traditional Technology for Environmental Conservation and Sustainable Development in the Asian-Pacific Regionpp: 171-175

43. Nayyar S(2001) Inside the mind of Gen Y. American Demographics 23: 6.

44. Cui Y, Trent ES, Sullivan PM, Matiru GN(2003) Cause-related marketing; how generation Y responds. International Journal of Retail & Distribution Management 31: 310-320.

45. Saunders M, Lewis P, Thornhill A(2009)Research methods for business student. 5th ed. Essex: Pearson Education Limited.

46. Sarker S, Lau F, Sahay S(2001) Using an Adapted Grounded Theory Approach for Inductive Theory Building About Virtual Team

Development. The DATA BASE for Advances in Information Systems 32: 39-56.

47. Kosulwat V(2002) The Nutrition and Health Transition in Thailand. Public Health Nutrition 5: 183-189.

48. Shannon R (2008) Thailand's Battle Between Modern and Traditional Trade Food Retailers: Interview with BoonchaiChokwattana, President and CEO of The SahaPatanapibul Plc. The Retail Digestpp: 38-43.

49. Euromonitor International (2010)Thailand Income & Expenditure.

50. Thompson AK, Moughan PJ(2008) Innovation in the Foods Industry: Functional Foods. Innovation: Management Policy & Practice 10: 61-73.

51. RahaviEB ,Kapsak WR(2010) Health and Wellness Product Development. Prepared Foodspp: 19-28.

52. Scarpa R, Thiene M, Marangon F(2007)The value of collective reputation for environmentally-friendly production methods: The case of val di gresta. Journal of Agricultural & Food Industrial Organization 5.

53. Mooij MD(2004)Consumer Behavior andCulture: Consequences for Global Marketing and Advertising. London: SAGE Publications Limited.

54. Shannon R, Lockshin L(2001) Exploring Consumer Trial of Private Label Brands in Thailand. Proceedings of the ANZMAC Conference.

55. Sorina-Raula G, Liviu C, Georgeta-Madalina M(2008)The Role of Advertising in the Purchase Decision Process. Annals of the University of Oradea, Economic Science Series 17: 895-900.

56. LengCY ,Botelho D(2010) How Does National Culture Impact on Consumers' Decision-making Styles? A Cross Cultural Study in Brazil, the United States and Japan. Brazilian Administration Review 7: 260-275.

57. Peter JP, Olson JC(2010)Consumer Behavior& Marketing Strategy. 9thed. New York: McGraw-Hill Companies.

7

CORPORATE SOCIAL RESPONSIBILITY REPORTING IN THE HOTEL INDUSTRY

Maria Krambia-Kapardis

Department of Hotel and Tourism Management,
Cyprus University of Technology

ABSTRACT

Global trends in the field of Corporate Social Responsibility (CSR) such as the welfare of employees and the shortage of natural resources, along with technological innovations have all resulted in a new set of stakeholder requirements. Stakeholders are nowadays demanding not only for sustainable products and/or services, but also require the corporations behind the brands they are consuming to offer greater transparency and reveal their accountability with regards to the way they carry out business. Consequently, to meet the needs of stakeholders, many corporations have already incorporated the sustainability discourse in their business strategies. More recently, corporations have started to publish sustainability reports, as a means of informing their stakeholders and shareholders about their organizations' environmental and social performance.

Previous studies within the hospitality sector, indeed confirm that hoteliers have already started to incorporate CSR and the reporting practise in their businesses, as they have documented an increase in sustainability reporting in the hotel sector over the past few years. Evidence suggests that by incorporating sustainability in their business strategies, hotels can decrease their operating costs, differentiate their product, and enhance their brands' image. In addition, by drafting sustainability reports to comment upon these issues, hotels can attract more customers, provide better hospitality products, assist in the formation of sustainable tourism destinations, and improve their overall performance efficiency. The paper reaches the conclusion that sustainability reporting within the hotel sector will gain more momentum in the near future as it will become a valuable tool for gaining a competitive advantage

With this in mind, the purpose of the paper is to explore the notion of CSR within the hotel industry and to document the benefits derived by the reporting practise. For the purposes of the paper, the researchers will review existing literature in the field of CSR, sustainability and sustainability reporting in the hotel industry and retrieve data regarding the environmental and social performance from the hotels' corporate websites.

KEYWORDS

corporate social responsibility, sustainability, sustainability reporting, hotel industry

1. INTRODUCTION

Global trends in the field of Corporate Social Responsibility (CSR) such as the welfare of employees and the shortage of natural resources, along with technological innovations have all resulted in a new set of stakeholder requirements. Stakeholders nowadays demand not only environmental protection and sustainable products and/or services, but also require businesses to offer greater transparency and accountability with regards to the way they carry out business. As a result, in an effort to satisfy the demand for sustainability, many business organisations are changing the way they operate and create profits, whereas they are also publishing sustainability reports in order to inform both their shareholders and stakeholders regarding their organisations' environmental and social performance.

The hotel and tourism industry, being amongst the world's fastest growing sectors (De Grosbois, 2012) with a 9% contribution to the global GDP, amounting to US $6.6 trillion (World Travel & Tourism Council, 2013 report) cannot afford to stay affluent upon the new stakeholder requirements. According to a Deloitte survey, 34 percent of the surveyed travellers are now seeking for environmentally friendly hotels (Deloitte, 2008 in Millar and Baloglu, 2011). Mindful of the stakeholder's environmental concerns, hoteliers have started to incorporate responsible business practices into their businesses (Millar and Baloglu, 2011:302) whereas many hotel service providers have already committed in adopting the sustainability reporting practice in an effort to cover the demand for greater transparency (Dwyer, 2005; Assaf et al., 2012).

With this in mind, the purpose of the paper is to explore the notion of CSR within the hotel industry and to document the benefits derived by the reporting practise. For the purposes of the paper, the researchers will review existing literature in the field of CSR, sustainability and sustainability reporting in the hotel industry and retrieve data regarding the environmental and social performance from the hotels' corporate websites.

2. THE NEED FOR SUSTAINABILITY

2.1 Corporate Social Responsibility

The notion of Corporate Social Responsibility (CSR) refers to the responsible manner by which an organisation can showcase its commitment to carrying out business and its daily operations on an ethical basis. It provides a framework by which business organisations can do well financially by doing good socially and environmentally, and it includes "actions that appear to further some social good, beyond the interests of the firm and that which is required by law" (McWilliams and Siegel, 2001:117). CSR actions mainly focus on the protection, conservation and restoration of the natural environment, the ethical treatment of employees and consumers, and the betterment of the society through community investment.

With the awareness growing regarding the social and environmental consequences of business activity, stakeholders such as governments, NGOs, customers and the public are nowadays demanding businesses to implement responsible business practises (de Grosbois, 2012). As a result, over the last two decades, the notion of CSR has become the new way of doing business and has gained increased popularity, as it has been implemented by a plethora of business organizations as a means of addressing their social and environmental responsibilities by adjusting their business practises to lessen their negative impact upon the environment and the wider society and positively contribute in the communities in which they operate.

Over the past two decades, many terms have been devised in order to describe the multidimensional character of businesses which adopt ethical and sustainable behaviour towards their stakeholders and the wider society. Amongst the most popular terms for the social responsibility of businesses are: corporate sustainability, corporate citizenship and corporate environmental responsibility. Emphasizing on the broad scope of CSR, which goes beyond philanthropy to operating sustainably in all aspects of business activity, the World Business Council for Sustainable Development (WBCSD) has defined CSR as "the continuing commitment by business to behave ethically and contribute to economic development while improving the quality of life in the workforce and their families as well as the local community and society at large" (WBCSD, 1999: 3).

However, as there are no hard laws on implementing CSR and no specific frameworks to evaluate it, the business organizations that go over and above their legal obligations on employing CSR tend to inform the public regarding their CSR initiatives through voluntary communication practises (de Grosbois, 2012). Thus, given the voluntary character of CSR, in the 2001 Green book published by the EU Commission, CSR was defined as "a concept whereby companies integrate social and environmental concerns in their business operations and their interactions with their stakeholders on a voluntary basis" by which they can create new opportunities and gain a competitive advantage.

2.2 Sustainability in the hospitality industry

Whereas the business case for CSR by which businesses can create long term shareholder value through the better management of risks deriving from the economic, environmental and social conditions is still developing, the hospitality industry has started to firmly embrace the notion of sustainability. In the 1987 Brundtland report, sustainability was given a holistic and future oriented definition, defined as "a development that meets the needs of the present without compromising the ability of the future generations to meet their own needs" (WCED, 1987: 45).

Business organisations are influenced by the systems surrounding them, which include the business, stakeholder and global environment, and they also have an influence on these systems themselves. As business organisations are inherent parts of social systems and ecosystems, they have an obligation to become more socially and environmentally responsible and contribute to sustainable development. Consequently, the underlying principle of sustainability, which is often referred to as the "Triple Bottom Line" is comprised by three pillars or the "3 E's" Environment, Equity, and Economics which are seen as equal pillars in becoming a good corporate citizen. According to Assaf et al. (2012: 596), the Triple Bottom Line "reflects the firm's commitment to create financial, social and environmental values" on a balanced basis.

Many studies have proven a positive correlation between CSR and sustainability with profitability. Consequently, by addressing environmental and social impacts in a systematic way, even more hotel service providers could in fact be taking a leap towards corporate sustainability. According to an MIT study two thirds out of the 4000 survey manager respondents said that sustainability is necessary to being competitive in today's marketplace (MIT, 2012). Although the benefits from operating in a sustainable manner are mainly intangible, (ie. Competitive advantage, enhanced brand reputation, product/services innovation), business organisations are profiting from sustainability activities. Realising the potential benefits and having acknowledged the notion that the relevant stakeholders for their businesses are not only their customers and their shareholders, but also the external society and the natural environment, a growing number of hotel service providers nowadays choose to implement the sustainability discourse in their business strategies.

2.3 Sustainability reporting and its value

According to Morsing (2006:171), CSR communication is defined as the "communication that is designed and distributed by the company itself about its CSR efforts" and can be distributed through sustainability reports, advertising campaigns, the media, or the Internet (De Grosbois, 2012). A lot of studies in the field of sustainability have proved that businesses who manage sustainability issues are financially more successful (Gaggl, 2013). However, in order for consumers and stakeholders to be informed about the sustainability efforts of businesses, companies should publish their sustainability performance.

Gaggl, (2013) notes that , by drafting annual sustainability reports to comment on such issues, business organisations can enjoy:

a) enhanced reputation and brand image for being more transparent in the provision of non-financial information

b) better client relationships as the embedment of sustainability in all aspect of business will essentially stimulate innovation which in turn will create new products and/or services and reach new markets

c) better investors/stakeholders relationships as they will be provided with holistic information that will create long term value for the business

Enhanced reputation for being a 'good' employer, thus, expectations for employees will have high-potentials thereby attracting top of the line staff.

3. THE IMPORTANCE OF SUSTAINABILITY REPORTING IN THE HOTEL INDUSTRY

Businesses are recognizing the increasing need of sustainability reporting due to pressure exerted by stakeholders and consumers Existing research has established that stakeholders are increasingly seeking for greater information and transparency from the corporations behind the brands, services and products they consume (De Grosbois, 2012). Consequently, as the provision of financial information is not enough anymore to embrace all value drivers, CSR communication and sustainability reporting has been gaining momentum.

Being a significant sector within the tourism industry, the hotel industry has a crucial role in contributing to sustainability. Although being a huge provider of social and economic benefits, the industry has a severe impact on the natural environment. Thus, due to the impact of hotels upon the environment with regards to resource consumption, the United Nations Agenda 21 includes a chapter that focuses on sustainable tourism, and has provided the industry with a comprehensive blueprint for action to be undertaken on a global scale (Division for Sustainable Development 1997 in Hsieh 2012). Agenda 21, encourages business organisations "to report annually on their environmental records, as well as on their use of energy and natural resources" (United Nations, 1992). As a result, a plethora of guidelines and standards (ie., EMAS, ISO 50001, ISO26000, GRI) have been developed, all aiming to assist organizations, including hotels, in reporting their performances while minimising their environmental footprint. A study by PwC asserts that over a dozen major hotel operators, including Four Seasons, Hilton, Hyatt, InterContinental, Marriott, Taj and Starwood, are now employing green initiatives to assist them in reaching the minimum standards for a sustainable hotel (PwC, 2008).

Given the fact that previous studies have proven that sustainability reporting has a significant impact on hotel performance (Assaf et al., 2012), the move towards sustainability reporting is justified by the hotel's need to generate economic welfare and increased profits. However, a crucial differentiation factor for hotels to enable them to derive all the benefits from operating in a sustainable manner lies on how well they communicate their sustainability efforts. Epstein (2008) stated that in addition to the financial reports, the hotel industry must also provide communication with regards to their environmental commitment, whereas, Charkham (2005) noted that adequate flow of information is crucial for businesses' to be able to reveal their commitment to sustainability.

One way of making sure that the sustainability report fully addresses the stakeholder concerns is to base the report on the company's Corporate Social Responsibility Strategy, its different pillars and the corresponding action plans for each pillar. Through this process, there is consistency between strategy, actions and performance measurement and companies are able to focus resources on the most important issues for tracking, reporting, and improvement initiatives (Ioannou, 2012).

3.1 The rise of the green traveller

The rising ethical concerns regarding the effect that modern consumption has on the environment, the increased attention that environmental and social issues receive from the mainstream media, along with the ever-increasing availability of ethical products and services, have triggered the awareness of consumers regarding the impacts of their purchasing behaviour (Carrigan and Attalla, 2001; Connolly and Shaw, 2006; Crane and Matten, 2004). According to a study by Vora (2007), 43 million U.S. travellers expressed environmental concerns whereas the 2008 National Leisure Travel Monitor survey also found that 85 percent of leisure travellers consider themselves as environmentally conscious (Crocker 2008 in Millar and Baloglu, 2011). In addition to that, the Travel Foundation found that only 4 per cent of UK holidaymakers would choose a five-star hotel rather than a four-star hotel with better sustainability performance (Travel Foundation in PwC, 2008). Therefore, due to the fact that "tourists have become increasingly demanding", they "force hoteliers to adapt to their new tastes and preferences, including greater respect towards the environment" (Molina-Azorin et al. 2009: 517).

Consequently, by incorporating sustainability in their business strategies, hotels are likely to improve their guests' perceptions regarding the environmental quality of the hotel and the tourism product they offer, thereby obtaining differentiation from competitors (Molina-Azorin et al., 2009; Chan and Wong, 2006; Kirk, 1998). A study by Travelocity, has found that 80 per cent of travellers would be willing to pay extra so they can visit an eco-friendly destination (Travelocity in PwC, 2008). As a result of the increasing stakeholder demand for sustainability and the considerable interest in social and environmental performance information, many hoteliers are nowadays publicly reporting their social and environmental performance and financial outcomes (Jose and Lee, 2007; Hsieh, 2012). Many hotel websites are becoming an electronic platform to disseminate this kind of information and reach stakeholders quickly and at little cost. By responding to the public's demand for sustainability and accountability hotels can develop a positive image, increase employee loyalty while also attract and retain customers. One can therefore say that the driving force behind the hotels' decision to incorporate sustainability and report accountability is the rise of the green traveller.

3.2 Reporting creates value for the hotel industry

Legrand et al. (2012a) stated that the main objectives behind the incorporation of sustainability in the business strategy of hotels are the decrease of operating costs and the building of a positive brand image. While the hotel industry is neither responsible for severely polluting the environment, nor is accountable for consuming vast amounts of non-renewable resources, the industry has a significant effect upon the use of global resources (Kirk, 1995; Chan and Wong, 2006). The vast consumption of energy and water, along with the use of consumable products and the generation of solid and hazardous waste, all have a significant impact on the environment. The costs associated with these impacts are great, thus, hotel service providers have reduced revenues, increased operating and employee costs. However, through the use of strategic sustainability and reporting, businesses can track their footprint, minimize risks and increase their opportunities thereby reducing their costs and enhance their value. This is supported by Willy Legrand who found that 74 per cent of hotels are now investing in energy savings to reduce costs (Legrand in The Guardian, 2013).

International hotel chains are nowadays making CSR the centrepiece of their business strategies, underpinning environmental and social concerns within corporate decision-making. For instance, Wyndham Hotels & Resorts (2009) see CSR as "a way of living, working and playing that embodies our vision and values, celebrates our diversity and supports a balance of professional and personal needs". By reporting their environmental performance and by communicating the initiatives and objectives of their CSR strategies, businesses in the hotel industry cannot only reach the niche market segment of green travelers, but they can also reinforce the atmosphere of partnership and fair treatment within the industry (Bohdanowicz et al. 2011). As a result, hotels have the opportunity to enhance their employees' job satisfaction (Spector, 1997) and organisational commitment, while also attract and retain skillful employees, thereby creating long-term value for their organisations.

3.3 Reporting creates value to the destination

The hotel industry has a vested interest in protecting the environment as its success heavily depends on the attractiveness of its natural surroundings (Chan and Wong, 2006). Since many hotels are located in major cities and are often located near cultural or natural heritage sites, by attracting travellers they impose an ecological footprint (Kirk, 1995). However, given that an unspoilt environment is a vital for the attractiveness of any tourist destination, hotels must ensure the long-term environmental sustainability of tourist activity. Thus, by adopting sustainable business practices and by reporting their footprint, hotels can monitor their performance; reduce their impact upon the environment across the value chain, thereby assisting destinations to become more sustainable.

The World Tourism Organization (1981) has defined sustainable tourism as "tourism which leads to management of all resources in such a way that economic, social and aesthetic needs can be filled while maintaining cultural integrity, essentials ecological processes, biological diversity and life support systems". Given the fact that the ultimate goal of the hospitality industry is to shift unsustainable tourism development to a more favoured position hotels are nowadays choosing to report their footprint impact next to their financial results. A report by the United Nations Environment Program (2011) has identified tourism to be amongst the ten economic sectors which are the driving force behind the transition to a green economy, which aims to reduce environmental risks and ecological scarcities and increase human well-being and social equity.

By reporting on the triple bottom line, the hotel industry can measure and monitor the economic, social and environmental impacts generated by tourism. The OECD's Toward Green Growth report (2012) suggests that market instruments, such as sustainability reporting, boost green innovation and address the externalities associated with environmental challenges. Through the reporting practice, hotels can adopt and develop green innovations which will enhance their resource efficiency and lower the costs of addressing environmental challenges. Measuring thus a destination's footprint, will in turn result in the development of a decision-making framework aiming to achieve sustainable development in any given destination. As a result, by reporting sustainability, hotels and other facilities in the hospitality value chain can create value to the destination itself.

4. CONCLUSIONS

The rise of CSR, which suggests that business should become more sustainable by minimising their environmental impact and more accountable towards their stakeholders, is confirmed by the fact that nearly all international companies have CSR policies in place (Franklin, 2008). Being amongst the world's fastest growing sectors (de Grosbois, 2012) and given that the conservation of natural surroundings is a vital factor for their success, the hotel industry could not afford to stay affluent upon these new demands. Thus, after having acknowledged the value of the sustainability discourse in the corporate strategies of business organisations world-wide, it is no surprise that several studies have indeed reported an increase in sustainability reporting throughout the hotel industry (Dwyer, 2005; Assaf et al. 2012; Legrand et al., 2012a; de Grosbois, 2012).

In principle, hotels, have many viable opportunities to engage in environmental sustainability. The main drivers in adopting transparent business practices which focus on ethical values, social equity and environmental conservation, are the growing stakeholders' demands, the decrease of operating costs, the differentiation of products and the enhancement of the brands' image. As a result, issues pertaining the implementation and reporting of sustainability are very relevant, timely and critical for the sustainability of the tourism industry (de Grosbois, 2012) in the critical financial times most countries are experiencing

Due to the higher-than-normal consumption of resources and the negative impact upon the environment, hotels have strong economic incentives to employ sustainable business strategies and to communicate their environmental performance to both their shareholders and stakeholders. By drafting sustainability reports, hotels can offer greater transparency and reveal their accountability through their environmental and social performance results. In addition to that, the effort to report can promote and create value not only for businesses in the hotels sector, but also for the destination itself, as it will lead to increased tourism innovation and therefore better hospitality products. It is therefore suggested that the sustainability reporting practise in the hotel industry will grow even more in the near future and will become a valuable tool for hotels worldwide for gaining a competitive advantage and improving performance levels.

REFERENCES

1. Assaf, A. G., Josiassen, A. & Cvelbar, L. K (2012), 'Does triple line reporting improve hotel performance?'. *International Journal of Hospitality Management,* 31(2), pp. 596-600.

2. Barnes, P. E. (1996), 'Green Standards'. *B & E Review,* October–December, 24-28.

3. Bohdanowicz, P., Zientara, P. &Novotna, E. (2011) 'International hotel chains and environmental protection: an analysis of Hilton's

we care! Programme', *Journal of Sustainable Tourism.* 19(7), pp. 797-816.

4. Braw, E. (2013), 'Sustainable tourism: the rise of the eco-boutique hotels', *The Guardian.* Retrieved February 10, 2014, from:http://www.theguardian.com/sustainable-business/blog/sustainable-tourism-eco-boutique-hotels
5. Carrigan, M. & Atalla, A. (2001), 'The myth of the ethical consumer—Do ethics matter in purchase behavior?', *Journal of Consumer Marketing*, 18 (7), pp. 560-577.
6. Chan, E.S.W. & Wong, S.C.K. (2006), 'Motivations for ISO 14001 in the hotel industry', *Tourism Management.* 27(3), pp. 481–492.
7. Chan, E.S.W. (2007), 'Environmental measures for hotels' environmental management systems ISO14001', *International Journal of Contemporary Hospitality Management,* 21(5), pp. 542-560.
8. Charkham, J. (2005), *Keeping better company: corporate governance ten years on,* Oxford: Oxford University Press
9. Christensen, P. & Rasmussen, B. D. (1998), 'Implementation of environmental management in the Danish printing house industry', *Eco-Management and Auditing,* 5(2), pp. 88–94.
10. Connolly, J. & Shaw D. (2006), 'Identifying Fair Trade in Consumption Choice', *Journal Of Strategic Marketing,* 14 (4), pp. 353-368.
11. Crane, A. & Matten, D. (2004), *Business Ethics: A European Perspective,* Oxford: Oxford University Press.
12. De Grosbois, D. (2012), 'Corporate Social Responsibility reporting by the global hotel industry: Commitment Initiatives and performance', *International journal of Hospitality Management,* 31(3), pp. 896-905.

13. Dwyer, L. (2005), 'Relevance of triple bottom line reporting to achievement of sustainable tourism: a scoping study', *Tourism Review International,*9(1), pp. 79-94.

14. Epstein, M. (2008), *Making sustainability work,* Sheffield: Greenleaf Publishing.

15. European Commission. (2001), *Green Paper: Promoting a European Framework for Corporate Social Responsibility,* Retrieved December 20, 2013, from http://eur-lex.europa.eu/LexUriServ/site/en/com/2001/com2001_0366en01.pdf

16. Franklin, D. (2008), Just good business, *Economist,* 8563, 3-22.

17. Gaggl, P. (2013), 'New approaches in Sustainability Reporting: From Theory to Practise', *Sustainability reporting training PwC.* Presentation given on 4-7 November at Cyprus University of Technology, Limassol, Cyprus.

18. Global Reporting Initiative (2011), *'Sustainability reporting guidelines',* Retrieved December 24, 2013, fromhttps:// www.globalreporting.org/resourcelibrary/G3.1-Guidelines-Incl-Technical-Protocol.pdf

19. Global Reporting Initiative (2011), *'About GRI',* Retrieved December 18, 2013, from https://www.globalreporting.org/Information/about-gri/Pages/default.aspx

20. Hersey, K. (1998), 'A close look at ISO14000', *American Society of Safety Engineers,* 43(7), pp. 26-29

21. Holcomb, J. L., Upchurch, R. S., & Okumus, F. (2007), 'Corporate social responsibility: what are top hotel companies reporting?', *International Journal of Contemporary Hospitality Management,* 19(6), pp. 461-475.

22. Hooper, P. D. & Greenall, A. (2005). 'Exploring the potential for environmental performance benchmarking in the airline sector', *Benchmarking: An International Journal,* 12(2), pp. 151-165.

23. Hsieh, Y.C. (2012), 'Hotel companies' environmental policies and practices a content analysis of their web pages', *International Journal of Contemporary Hospitality Management,* 24(1), pp. 97-121.

24. Ioannou G.A. (2012), 'Corporate responsibility reporting', *PwC Cyprus, Accounting & Reporting Newsletter*, February 2012.

25. International Organization for Standardization (2010), *'Social responsibility - Discovering ISO 26000'*. Retrieved December 18, 2013, from:*http://www.iso.org/iso/discovering_iso_26000.pdf*

26. Iwanowski, K. & Rushmore, C. (1994), 'Introducing the eco-friendly hotel', *The Cornell Hotel and Restaurant Administration Quarterly*, 35(1), pp. 34-38.

27. Jose, A. & Lee, S. (2007), 'Environmental reporting of global corporations: a content analysis based on website disclosure', *Journal of Business Ethics.*4(3), pp. 307-321.

28. Kirk, D. (1995). 'Environmental management in hotels', *International Journal of Contemporary Hospitality Management*, 7(6), pp. 3–8.

29. Legrand, W., Huegel, E.B., & Sloan W. (2012b), 'Learning from best practices: Sustainability reporting in international Hotel Chains', *Advances in Hospitality and Leisure*, 9, pp. 119-134.

30. Legrand, W., Kirsche, C., Sloan, W., & Simons-Kaufman, C. (2012a), 'Making 20 2020 happen: is the hospitality industry mitigating its environmental impacs? The barriers and motivators that German hoteliers have to invest in sustainable management strategies and technologies and their perceptions of online self help toolkits'. In F. Pineda & C.A. Brebbia (Eds.) *Sustainable Tourism V*, (pp. 115-126), Southampton: Wessex Institute of Technology.

31. McWilliams, A. & Siegel, D. (2001), 'Corporate social responsibility: A theory of the firm perspective', *Academy of Management Review*, 26(1), pp. 117-127.

32. Millar, M. & Baloglu, S. (2011), 'Hotel guest's preferences for green guest room attributes', *Cornell Hospitality Quarterly*, 52(3), pp. 302-311.

33. MIT Sloan Management Review & The Boston Consulting Group. (2012), *'Research report winter 2012 Sustainability nears a tipping point',* Retrieved December 19, 2013, from: http://c4168694. r94. cf2.rackcdn.com/MIT-SMR-BCG-Sustainability-Nears-a-Tipping-Point-Winter-2012.pdf

34. Molina-Azorin, J.F., Claver-Cortes, E., Pereira-Moliner, J., & Tari, J.J. (2009), 'Environmental practices and firm performance: an empirical analysis in the Spanish hotel industry', *Journal of cleaner production.* 17(5), pp. 516-524.

35. Morsing, M. (2006), 'Corporate social responsibility as strategic auto-communication: on the role of external stakeholders for member identification', *Business Ethics: A European Review.* 15(2), pp. 171-172.

36. OECD Tourism Papers (2012), *'Green Innovation in Tourism Services'.* Retrieved February 18, 2014, from:http://www.oecd. org/cfe/tourism/GREEN%20INNOVATION%20IN%20TOURISM%20WORKING%20PAPER.pdf

37. Price Waterhouse Coopers (2008), *'Here to stay – Sustainability in the travel and leisure sector',* Retrieved February 10, 2014, from: https://www.pwc.co.uk/assets/pdf/here-to-stay-sustainability.pdf

38. Spector, P.E. (1997), *Job satisfaction: Application, assessment, cause and consequences,* Sage: Thousand Oaks, CA.

39. United Nations Environment Programme (2011), *'Towards a Green Economy',* Retrieved February 22, 2014, from: http://www. ipu.org/splz-e/rio+20/rpt-unep.pdf

40. United Nations. (1992), *'United Nations conference on environment & development: Agenda 21',* Retrieved December 18, 2013, from:http:

//sustainabledevelopment.un.org/content/documents/Agenda21.pdf

41. Vora, S. (2007), *'Business travellers go green',* Retrieved February 20, 2014, from: http://www.forbes.com/2007/06/21/business-travel-environment-forbeslife-travel_cx_sv_0622green.html

42. World Business Council for Sustainable Development (1999), *'Corporate Social Responsibility: Meeting changing expectations',* Retrieved December 23, 2013, from:http://oldwww. wbcsd. org/DocRoot/hbdf19Txhmk3kDxBQDWW/CSRmeeting.pdfhttp://oldwww.wbcsd.org/DocRoot/hbdf19Txhmk3kDxBQDWW/CSRmeeting.pdf

43. World Commission on Environment and Development (1987), *'Our Common Future: Brundtland Report 1987',* Retrieved December 20, 2013, from:http://conspect.nl/pdf/Our_Common_Future-Brundtland_Report_1987.pdf

44. World Tourism Organization (1981), *Saturation of Tourist Destinations: Report of the Secretary General*, Madrid: WTO Publications.

45. World Travel & Tourism Council (2013), *'Travel & Tourism Economic Impact 2013 World'*, Retrieved December 19, 2013, from: http://www.wttc.org/site_media/uploads/downloads/world2013_1.pdf

46. Wyndham Hotels & Resorts (2009), *'WyndhamGreen',* Retrieved February 18, 2014, from: http://www.wyndhamworldwide. com/corporate-responsibility/wyndham-green/green-companies

8

DIMENSIONS OF E-CRM: AN EMPIRICAL STUDY ON HOTELS' WEB SITES

Noor Raihan Ab Hamid1, Aw Yoke Cheng2 and Romiza Md Akhir1

1Faculty of Business Administration, University of Management and Technology (UMTECH), Petaling Jaya, Selangor, Malaysia

2Faculty of Business and Management, Asia Pacific University College of Technology and Innovation (UCTI), Technology Park Malaysia, Kuala Lumpur, Malaysia

ABSTRACT

Electronic consumer relationship management (E-CRM) features are among the main factors contributing to the success of any ecommerce Web sites. E-CRM programmes and implementation should focus on effectiveness of managing consumer relations on the Internet through effective use of Internet features either to deliver updated and accurate information, offering customer service support as well as providing personalized and value-added services. Thus, the aims of this research are to identify E-CRM dimensions and perceived important features from a consumers' perspective. The primary data for this research were collected through a questionnaire survey. To determine the dimensions of E-CRM, a factor analysis was performed. The findings reported 14 dimensions of E-CRM which were perceived as important elements affording consumers repeated visits and purchases. The limitations of the study and future directions of research are discussed herein.

KEYWORDS

Electronic Customer Relationship Management, Hotels, Web sites, Retention.

1. INTRODUCTION

The business environment of the digital economy has evolved into competition which is based on speed, ability to innovate and forging strategic collaboration with industry players as well as consumers. The advent of advance technology such as the Internet and mobile applications enables companies, large or small to compete in a cost effective manner than ever before. In a service sector, where efficient coordination is found, information and communication are critical, the low-cost-highly- efficient Internet technology allows value adding and even personalized services to be delivered. Primarily, competition in hospitality and tourism industry is centred on delivering consumer value through quality services to domestic and international tourists. Developing an effective consumer relationship via technology platform such as Internet may assist in increased consumers' satisfaction, loyalty and retention. Indeed, consumer relationship management (CRM) has been regarded as one of the most important concepts in modern marketing. By far, managing consumer relationship with the advancement of Web 2.0 requires businesses to collaborate and engage with their consumers through purposeful use of technology, such as social media.

Malaysian markets have recently begun to understand the importance of electronic consumer relationship management (E-CRM) (Ab Hamid, 2008; Loh 2000). The interest was led by traditional consumer management economics. It costs the industry five times as much to acquire a new consumer than to retain an existing one; therefore, building long-term and sustainable consumer relationships was a good strategy, from the point of view of both building life-time value relationships and business cost savings involved (Peppers and Rogers 1996). Khatibi et. al., (2002) found that there is a positive trend of increasing awareness of the benefits of Internet marketing. Factors, such as global access and ubiquity, convenience, increased product information and availability of value-added services on the Internet were cited as contributing factors of enhancing consumer satisfaction and loyalty in the cyberspace (Ab Hamid et al, 2010).

This research aims to identify the consumers preferred features and services of hotels Web sites leading to repeat visits and purchases. Premised on the belief that delighted consumers are likely to return and become loyal to service providers, this study adapted the E-CRM dimensions proposed by Ab Hamid and McGrath (2005) to empirically identify the dimensions and features of E-CRM relevant to the hotel industry. In an intense competition, it is crucial for hotel operators to understand the elements which are perceived as important in influencing consumers' repatronage behaviour. Hence, a study to ascertain the hotels Web sites service attributes and dimensions perceived important from consumers' evaluation is worthwhile. This research was carried out as a pilot study to determine the reliability and validity of the measurement scales used, which were adapted from E-CRM dimensions proposed by Ab Hamid and McGrath (2005).

2. RELATIONSHIP MARKETING AND THE HOTEL INDUSTRY

2.1 Growth of the Hotel Industry

Hotel service is one of the most important sectors of hospitality and tourism industry contributing significantly to the economic growth in Malaysia. According to Tourism Malaysia (2011), there were 24.6 million tourists who have visited Malaysia and in turn contributed with MYR 56.5 billion receipts to the country in year 2010 as compared to 23.6 million tourists and with MYR 53.4 billion receipts in the previous year. The arrivals were largely from the Association of South-East Asian Nations (ASEAN) countries which recorded 77.8%. Despite the gradual growth of tourist arrival to Malaysia, the overall hotel occupancy in Malaysia has recorded a slight reversal of growth in year 2010. Much of the decline was due to the recent global economic crisis most of which effect was felt in the early quarters of 2010. The hotel occupancy rates in year 2010 as compared to year 2009 has shown slight decline of 1.6 percent for the whole of Malaysia, however metropolitan areas such as the Klang Valley were not affected by this downturn. For example, in Kuala Lumpur, Selangor, and Putrajaya, the occupancy rates have increased 4.2 percent, 2.4 percent and 1.4 percent respectively (Tourism Malaysia, 2011).

Indeed the slight decline has called for serious attention from the hotel operators to rethink the importance of building tourist (consumer) relationship. One of the ways to strategize is to aggressively plan for relationship marketing through various marketing channels including the Internet.

2.2 Electronic Consumer Relationship Management (E-CRM)

According to Grönroos (1994) service firms by their inherent characteristics of inseparable production and consumption possess the elements necessary to forge relationships with consumers. Because most companies offer almost the same core product/services, differentiation will be of greater interest to companies with the strongest capability to develop long-term consumer relationships (Zineldin, 1999). Consumers perceived that service quality, satisfaction and value have played an important role in service management (traditionally) in the context of online businesses, so a holistic and multi-purpose view of the whole relationship is required (Holmlund & Strandvik, 1999; Luck & Lancaster, 2003).

In a study of the current use of the Internet as a marketing tool in the hotel industry, Gilbert et al. (1999) present an argument for the application of the RM model as a framework for the development of hotel Web sites. They contend that hotels need a framework that can bridge the gap between simply connecting to the Web and harnessing its power for competitive advantage. However, the pursuit of the marketing objective will be constrained by the possibilities presented within the hotel's market area, the hotel's location and its current product positioning (Jeffrey and Barden, 2000). WebProbe claims that if hoteliers know more about their visitors, they will be able to redesign their sites to help them build stronger relationships with them. Geoffrey Breeze, vice-president of marketing at Hilton International maintains that hotels have far more information about their guests than they can actually use (Caterer & Hotelkeeper, 2001).

Drawn from previous literature, this study investigated perceived important E-CRM dimensions and features in managing consumer relations. Ab Hamid and McGrath (2005) reported 12 dimensions of E-CRM program namely: information quality, ease of navigation, consumer service quality, fulfilment, integrated marketing channels, online community, rewards, personalization level, site security, value-added services, perceived trust and price attractiveness.

The following paragraph describes Ab Hamid and McGrath (2005) dimensions of E-CRM:-

a) Information Quality

Based on information systems literature, information quality is a reflection of relevancy, recency, sufficiency, consistency and understandability. Since a consumer's decision making efficiency improves when searching is simplified, information presented on the sites should be easy to understand and up-to-date.

b) Ease of Navigation

In brief, ease of navigation includes both good organization (format) of the content layout as well as simple-to-use navigation (ease of use). Sites that are easy to navigate offer quick access to information needed with minimized effort.

c) Consumer Service Quality

Consumers constantly demand for careful, continuous, useful communication with company representatives. Since these attributes are frequently identified as salient dimensions in store selection behaviour, company representatives should have the knowledge and basic technology skills to answer online questions.They should understand consumer-specific needs, have the capacity to handle problems that arise and address consumer complaints in a friendly manner.

d) Fulfilment

The order fulfilment process is concerned with delivering the right product at the right time and responding to consumer inquiries.

e) Integrated Marketing Channels

For click-and-mortar companies, integration of the virtual and physical operations is crucial to ensure continuous consumer relationships in both channels.This is because consumers want to be able to deal with companies as single entities.They also expect to get the same consumer service level via all channels.

f) Online Community

Consumers can exchange information with each other and obtain online help from their online members when a product- related problem arises. These experiences create stickiness – the extent to which a company is able to attract consumers to use a site relatively longer and return, and consumers are more likely to revisit the site.

g) Rewards

Reward is another attribute deemed important to attracting consumers to repatronize. Reward programs allow consumers to collect points for every purchase from or visit to a site, in exchange for free gifts, coupons or cash rebates. The program also provides greater membership privileges for returning consumers.

h) Personalization Level

Personalization attracts consumers to come back repetitively since it turns consumers into product 'makers' rather than simply product 'takers' (Winer 2001), Thus, consumers are empowered in ways that they can choose their own preferred design, colour, product updates and other attributes that go well with their tastes.

i) Site Security

Consumers are concerned about online payment security, reliability and privacy policy since they have to provide their personal details and credit card information in the ordering process. This concern increases the perception of risk and simultaneously reduces the level of trust in an online company, which in turn adversely affects satisfaction. Hence, online companies that clearly communicate to consumers on how their private and transaction data are secured are more likely to benefit from increased consumer satisfaction (Ab Hamid et al., 2010).

j) Value-added Services

Consumers will feel empowered and in control when they are able to easily find product/service, learn more about it and quickly make a purchase decision. For example, a site may provide features such as shopping carts, one-click ordering and order tracking to provide superior convenience and higher levels of consumer control.

k) Perceived Trust

Since online transactions are associated with perceived risk, trust and confidence in the firm seem imperative in a consumer's decision making. Assurance of security in using the Web can be done by implementing security features such as digital certificates, secure servers and third party trusting agencies.

l) Price Attractiveness

Online consumers can easily compare prices across e-tailers (retailers who use the Internet as their only marketing channel). Consequently, if a price paid is higher than what is found in other e-tailers for a similar product, the consumer is significantly less satisfied.

3. RESEARCH FRAMEWORK

3.1 Questionnaire Design

We designed the questionnaire used in this study based on the research objectives discussed.The questionnaire consisted of two parts namely Section A and B. The first part, Section A, consisted of consumers opinion about the features they perceive as important on hotels Web sites which will induce repeated visits or purchases. Section B contained demographic information such as respondent age group and income level.These questions included respondents access location, number of years using the Internet, types of Internet activities and time spent in a week on the Internet. A 5-point labelled Likert type scale was used with anchor from 1= "Most important" to 5= "Not important at all".

Ab Hamid and McGrath (2005) conducted a study on E-CRM on Web sites in general and reported 12 salient dimensions. It is deemed appropriate to adapt Ab Hamid and McGrath's (2005) model since most of the dimensions are commonly found on Web sites. Additionally, extending from Ab Hamid and McGrath's (2005) model, this research investigated other items which were deemed appropriate for the hotel industry such as reservation tracking, and current trend of the Internet, that is the use of social media features. Consequently, a total of 69 items were used to test consumers' opinion on perceived important features of E-CRM on hotels Web sites.

3.2 Sampling

This research served as a pilot test to assess the measurement scales in terms of their reliability and validity which will be used for further study of the subject matter. Primary data for this research were collected using personally-administered questionnaires as this method will ensure a higher response rate. Research assistants helped in contacting respondents via email, and distributing and collecting the questionnaire in person. Secondary data for this study came mainly from newspapers, journals, publications, magazines, and books accessed via the Internet.

The main source of users list came from various education, government and corporate institutions in Klang Valley, Penang, Johor Bahru and Kuching.The nature of work that people do in these institutions requires the use of Internet; hence they are appropriate source of primary data for this study. For example, it is common for university staff and students to use the internet for communication or to conduct research. Likewise, most companies in the selected cities have internet connections on their premises and the method of communication within these companies is mainly email. Thus, the choices of these institutions are deemed appropriate for this study. For the purpose of the pilot test, 200 respondents were chosen using purposive (judgment) sampling, as the sample should have experience in surfing hotels Web sites. A total of 162 (81%) responded to the survey while the remaining respondents could not be contacted or declined to participate. Descriptive analysis was used to report the profile of research respondents while exploratory factor analysis was conducted to examine the dimensions of E-CRM.

4. RESEARCH METHODS AND DESIGN

4.1 Data Analysis and Results

Frequency distributions were calculated for all individuals in this research and are summarised in Table 1. As indicated, the respondents were almost evenly split by gender (57 per cent male and 43 per cent female). Most of the respondents were 21 to 30 years of age (77 per cent), followed by the age groups of below 20 years and 31 to 40 at 10 per cent and 7 per cent, respectively. These three groups contributed approximately 87 per cent of the overall total in terms of age. 70 per cent of the respondents had spent at least 15 years in education and held a bachelor's degree. Most of the respondents were working adults (86 per cent) and earned between MYR 24 000 - 48 000 per annum (69 per cent).

Table 1: Respondent Profile

Items	No. of Respondents	%	Items	No. of Respondents	%
1. Gender			**4. Occupation**		
- Male	76	57	- Student	16	12
- Female	58	43	- Non Executive	3	2
			- Executive	112	84
2. Age					
- Below 20 years	14	10	- Pensioner	3	2
- 21-30 years	103	77	**5. Monthly income**		
- 31-40 years	9	7	- less than MYR 2 000	19	14
- 41-50 years	7	5	- MYR 2 000 – MYR 4 000	93	69
- Above 50 years	1	1	- MYR 4 001 -MYR 6 000	10	8
			- MYR 6 001 and above	12	9
3. Education level					
- SPM	5	4	**6. Race**		
- STPM/Dip/Matrices	25	19	- Malay	12	9
- Degree	94	70	- Chinese	48	36
- Masters	8	6	- Indian	12	9
- PhD	1	1	- Others	62	46
- Others	1	1			

4.2 Factor Analysis

A principal component analysis (PCA) was performed on the data set to reduce a large number of observed variables into a smaller number of factors measuring different constructs. 69 items were used to operationalize the variables and to discern the salient dimension of E-CRM under study. Based on the Eigenvalue of more than 1.0 a Varimax rotation yielded 14 dimensions of E-CRM for hotels Web sites namely ease of navigation; information quality; customer service efficiency; payment options; personalization level; rewards; channel integration; use of social media; security and privacy policy; multimedia features; reservation and tracking; lower rates; fulfilment of service and quality services.

Table 2 shows the list of E-CRM dimensions and the Cronbach's alpha values. The alpha values indicated that the scales used to measure the dimensions were acceptable (> 0.60).

Table 2. List of E-CRM Dimensions for Hotels Web Sites

Factors/Dimensions	Cronbach's Alpha Value
Information quality	0.64
Quality services	0.68
Ease of navigation	0.79
Consumer service efficiency	0.72
Payment options	0.64
Security and privacy policy	0.62
Multimedia features	0.69
Personalization level	0.66
Rewards	0.86
Channel integration	0.65
Reservation and tracking	0.79
Use of social media	0.87
Lower rates	0.77
Fulfilment of service	0.85

5. DISCUSSION AND CONCLUSION

In order to be competitive, hotel operators should strive for adding value in their offerings on hotels' Web sites. Providing features such as reservation and modification to reservation facility, tracking transactions history, allowing some degree of personalization, and understanding consumers' specific needs and preferences enhance consumers' perceived value, thus increase loyalty. Apparently, the use of social media is gaining popularity and the trends are showing more literate and younger consumers becoming keen users of the new marketing channel. This study shows that consumers viewed hotels Web sites sharing past consumers' reviews through Blogs and Facebook as one of the important elements leading to intention to revisit.

Besides, consumer service quality should be given a top priority since it is the first contact point for consumers to evaluate whether a firm is reliable or not. When one finds difficulty in communicating with the consumer service; that is either an enquiry is not attended to within a reasonable time or the representative appears to have insufficient knowledge, he/she would immediately make an adverse evaluation about a firm's service performance.In a highly competitive environment such as the Internet, this unpleasant experience (leading to dissatisfaction) would leave firms losing not only one potential consumer but perhaps many more, due to "word-of-mouth". Hence, firms are well advised to attend to consumer complaints efficiently, clearly display the links to consumer service and help desk, and to have a consumer service representative always available.

Next, the quality of information should be another important attribute that firms should be concerned with. Since Internet technology stems from computer technology, which is believed to process and deliver information at speed, the relevancy and accuracy of information seem to be compulsory criteria of a Web site. A site that displays information that has expired at a point of time may be viewed as inefficient. In addition, it is vital for hotel operators to accurately capture and deliver services within consumers' expectations. Failure to fulfil services as per consumers' expectations will result in not only consumers' frustration but will adversely affect consumers' assessment of firms' performance. Then, ease of site navigation is also vital. Generally, Internet consumers are impatient, hence loading speed of the site, simplicity of use and accessibility are critical.

In this study, respondents were asked to recollect their experience of hotels Web sites. More in depth studies could be carried out to investigate the use of E-CRM in industry- specific environment such as financial sector, entertainment, health, government and education sector since E-CRM may imply different meanings from one industry to another.The sample respondents from this study were mainly working adults at executive level that have higher purchasing power. Future studies should be conducted to assess the opinions of younger adults who seem to have greater dependence on technology and higher literacy. Since this group of consumers would form a different market segment in the near future and may have different perception of matters of importance, studies to understand their behaviour and expectations of service are worthwhile. This study is concerned with consumers' perceived important features of E-CRM entailing consumers repeated visits and purchases. Although companies are well advised to adopt an E-CRM strategy, its implementation may vary depending on the business scale. For example, budget hotels may not be able to fully utilize the potential of Internet technology due to constraints in resources. Hence, E-CRM implementation in various business scenario merits further investigation.

ACKNOWLEDGMENT

The authors would like to express gratitude to the Ministry of Higher Education (MOHE), Malaysia, for giving the researchers the opportunity to conduct the research successfully. The support has enabled the researchers to pursue their goals and meet the objectives of this research. Appreciation also goes to the Faculty of Business Administration of the University of Management and Technology, and the Faculty of Business and Management of the Asia Pacific University College of Technology and Innovation, in providing facilities during the conduct of this research.

REFERENCES

1. Ab Hamid, N. R. (2008). "Consumers' Behaviour Towards Internet Technology and Internet Marketing Tools," International Journal of Communications, 2(3), 195-204.

2. Ab Hamid, N. R, & Kassim, N. (2004). 'Internet Technology as a Tool in Managing Customer Relationships,' The Journal of American Academy of Business Cambridge, 4 (1 &2), 103-108.

3. Ab Hamid, N. R. & Khatibi, A. A. (2006). 'Perceived Risk and Users' Experience Influence on Internet Technology Adoption,' WSEAS Transactions on Systems, 5 (12), 2766-2773 .

4. Ab Hamid, N. R. & McGrath, G. M. (2005). 'The Diffusion of Internet's Interactivity on E-tail Web Sites: A Customer Relationship Model,' Communications of the International Information Management Association, 2, 45-70.

5. Ab Hamid, N. R., McGrath, G. M. & Khatibi, A. (2007). 'Retaining Online Consumers: Evidence from South East Asia,' WSEAS Transactions on Systems, 6(3), 541-548.

6. Ab Hamid, N. R, Md Akhir, R. & Mashudi, P. M. (2010). 'An Assessment of the Internet's Potential in Enhancing Customer Relations,' International Journal of Arts and Sciences, 3(12), 265-281.

7. Anton, J. & Hoeck, M. (2002). 'E-Business Customer Service,' The Anton Press, Santa Monica, CA.

8. Arnott, D. C. & Bridgewater, S. (2002). "Internet, Interaction and Implications for Marketing," Marketing Intelligence and Planning, 20 (2), 86-95.

9. Baloglu, S. & Pekcan, Y. (2005). 'The Web Design and Internet Site Marketing Practices of Upscale and Luxury Hotels in Turkey,' Tourism Management.

10. Bradshaw, D. & Brash, C. (2001). "Managing Customer Relationships in the e-business World: How to Personalize Computer Relationships for Increased Profitability," International Journal of Retail and Distribution Management, 29, 520-529.

11. Burger, F. (1997). 'Web-database Supported Tourist Information on the Web,' in Tjoa, A.M. (Eds), Conference Proceedings: Information and Communication Technologies in Tourism, Springer, Wien, 39-46.

12. Cai, L., Card, J. A. & Cole, S. T. (2004). "Content Delivery Performance of World Wide Web Sites of US Tour Operators Focusing on Destinations in China," Tourism Management, 25(2), 219-27.

13. Calongne, C. M. (2001). "Designing for Web Site Usability," Proceedings of the 7th Annual Consortium for Computing in Small Colleges, Central Plains Conference on The Journal of Computing in Small Colleges, 39-45.

14. Chen, I. J & Popovich, K. (2003). "Understanding Customer Relationship Management (CRM): People, Process and Technology," Journal of Business Process Management, 9(5), 672-688.

15. Chung, T., & Law, R. (2003). "Developing a Performance Indicator for Hotel Websites," International Journal of Hospitality Management, 22 (1), 119-25.

16. Deimezi, O. & Buhalis, D. (2003). 'E-tourism and Developments in Greece,' Information and Communication Technologies in Tourism, Springer Verlag, Wien, 39-48.

17. Feinberg, R., & Kadam, R. (2002). "E-CRM Web Service Attributes as Determinants of Customer Satisfaction with Retail Web Sites," International Journal of Service Industry Management, 13, 432-451.

18. Economic Planning Unit of the Prime Minister Office, Government of Malaysia (2011). "The Malaysian Economy in Figures 2010,": External Sectors, [Online] [Retrieved: February 28, 2011], http://www.epu.gov.my/c/document_library/get_file?uuid=4f4dea61-0446-4803-b52f-65e85a37c5fe&groupId=34492

19. Gatsiounis, I. (2010). "Local Chains Sitting Out Hotel and travel Boom in Malaysia," Forbes Asia Magazine, [Online] [Retrieved: February 28, 2011], http://www.forbes.com

20. Gratzer, M. & Winiwarter, W. (2003). "The Role of the Internet in the SME Hotel Sector in Austria," Lecture Notes in Computer Science, 27(13), 85-95.

21. Gurau, C. (2003). "Tailoring e-service Quality through CRM," Managing Service Quality, 13, 520-531.

22. Heichler, E. (1997). "Internet Lacks Content for Women," Computer World , 31(19), 17.

23. Heung, V.C.S., (2003). "Internet Usage by International Travelers: Reasons and Barriers," International Journal of Contemporary Hospitality Management , 15(7), 370-378.

24. Ho, J. (1997). 'Evaluating the World Wide Web: A Study of 1,000 Commercial Sites,' Journal of Computer Mediated Communication , 3(1), [Online] [Retrieved: February 28, 2011], www.ascusc. org./jcmc/vol3/issue1/ho.html

25. Huizingh, E. K. R. E.(2000). "The Content and Design of Web Sites: An Empirical Study," Information & Management , 37(3), 123-34.

26. Jeong, M. & Lambert, C. U. (2001). "Adaptation of an Information Quality Framework to Measure Customers' Behavioural Intentions to Use Lodging Web Sites," International Journal of Hospitality Management , 20(2), 129-46.

27. Kandampully, J. & Duddy, R. (1999). "Relationship Marketing: A Concept Beyond Primary Relationship," Marketing Intelligence & Planning, 17(7).

28. Khatibi, A. A, Ismail, H . & Thyagarajan, V. (2002). "What Drives Customer Loyalty: An Analysis from the Hotels Industry," Journal of Targeting, Measurement and Analysis for Marketing, 11(1), 34–44.

29. Law, R. & Hsu, C. H. (2005). "Customers' Perceptions on the Importance of Hotel Web Site Dimensions and Attributes," International Journal of Contemporary Hospitality Management, 17(6), 493-503.

30. Lee, J, Lee, J. & Feick, L. (2001). "The Impact of Switching Costs on the Customer Satisfaction–loyalty Link: Mobile Phone Service in France," Journal of Services Marketing, 15, 35–48.

31. Liang, K. & Law, R. (2003). "A Modified Functionality Performance Evaluation Model for Evaluating the Performance of China Based Hotel Web Sites," Journal of the Academy of Business and Economics, [Online] [Retrieved: February 28, 2011], www.findarticles.com.

32. Lodhia, S. (2004). "Corporate Environmental Reporting Media: A Case for the World Wide Web," Electronic Green Journal, 20, [Online] [Retrieved: February 28, 2011], http://egj.lib. uidaho. edu/egj20/lodhia1.html

33. Loh, C. J. (2000, September 6). 'Using CRM Tools to Attract Customers,' New Straits Times, Kuala Lumpur, 10.

34. Luck, D. & Lancaster, G. (2003). "E-CRM: Customer Relationship Marketing in the Hotel Industry," Managerial Auditing Journal , 18(3), 213-231.

35. Mahfouz, A. (2000). "Web Sites Interface Design: External and Internal Factors," in Reich, S., Anderson, K.M. (Eds), Open Hypermedia Systems and Structural Computing, Proceedings of the 6th International Workshop, OHS-6, 2nd International Workshop, SC-2, San Antonio, Texas, 30 May-4 June, 46.

36. Malaysian Science and Technology Information Centre. (2008). 'Malaysian Facts and Figures,' [Online] [Retrieved: February 5, 2009], http://www.mastic.gov.my

37. Morrison, A. M., Taylor, S., Morrison, A. J. & Morrison, A. D. (1999). "Marketing Small Hotels on the World Wide Web," Information Technology and Tourism, 2(2), 97-113.

38. Murphy, J., Forrest, E. J., Wotring, C. E. & Brymer, R. A. (1996). "Hotel Management and Marketing on the Internet," The Cornell Hotel and Restaurant Administration Quarterly, 37 (3), 70-82.

39. Ness, J. A., Schroeck, M. J., Letendre, R. A. & Douglas, W. J. (2001). 'The Role of ABM in Measuring Customer Value,' Strategic Finance, 82, (9), 32-37.

40. O'Connor, P. (2003). "On-line Pricing: An Analysis of Hotel-company Practices," Cornell Hotel and Restaurant Administration Quarterly, Vol. 44 (1), 88-96.

41. Parets, R. (2002, August). 'Getting the Word out: Profiting from Smart Internet Marketing isn't Just for the Big Boys,' Lodging Magazine, 37-38.

42. Rachman, Z. & Buchanan, J. (1999). Effective Tourism Websites, Part 1: Literature Review and Features Surveys [Online] [Retrieved: February 28, 2011], http://mngt.waikato.ac.nz/depts/mnss/john

43. Rachman, Z. M. & Richins, H. (1997). "The Status of New Zealand Tour Operator Web Sites," The Journal of Tourism Studies, 8 (2), 62-82.

44. Rosen, D. E. & Purinton, E. (2004). "Web Site Design: Viewing the Web as a Cognitive Landscape," Journal of Business Research, 57 (7), 787-94.

45. Sigala, M. (2003a).'Competing in the Virtual Marketspace: A Strategic Model for Developing e-commerce in the Hotel Industry,' International Journal of Hospitality Information Technology, 3 (1), 43-60.

46. Sigala, M. (2003b). "Developing and Benchmarking Internet Marketing Strategies in the Hotel Sector in Greece," Journal of Hospitality & Tourism Research, 27(4), 375-401.

47. Sharif, R. (2004a, April 7). 'Malaysian Surfers: Online Banking is Hot, Shopping So-so,' The Star Online, [Online] [Retrieved: December 4, 2008], http://thestar.com.my.

48. Sharif, R. (2004b, April 27). 'IDC: Strong IT Spending this Year,' The Star Online, [Online] [Retrieved: December 4, 2008], http://thestar.com.my.

49. Sharif, R. (2004c, August 30). 'Online Banking Getting more Popular,' The Star Online, [Online] [Retrieved: December 4, 2008], http://thestar.com.my.

50. Suki, N. B. M., Ahmad, M. I. & Thyagarajan, V. (2002). "Motivation and Concern Factors for Internet Shopping: A Malaysian Perspective," The Electronic Journal for E-commerce Tools and Applications, 1, 1-18.

51. The Star Online. (2004, November 26). 'IT investment: Promising outlook for Malaysia,' [Online] [Retrieved: December 4, 2008], http://thestar.com.my.

52. Tourism Malaysia. (2011). "Facts and Figures 2010," [Online] [Retrieved: February 28, 2011], http://corporate.tourism.gov.my/research.asp?page=facts_figures

53. Tourism Malaysia. (2011). 'Malaysia Average Hotel Occupancy Rates,' Tourism Malaysia, [Online] [Retrieved: March 31, 2011], http://corporate.tourism.gov.my/tourismmalaysia_corpx/rpt3_hoteloccupancy.cfm?rpt=3

54. Van der Merwe, R. & Bekker, J. (2003). "A Framework and Methodology for Evaluating e-commerce Web Sites," Internet Research: Electronic Networking Application and Policy, Vol. 13 No.5, pp.330-41.

55. Van Hoof, H., Ruys, H. & Combrink, T. (1999). "Global Hoteliers and the Internet: Use and Perceptions," International Journal of Hospitality Information Technology, 1 (1), 45-61.

56. Wan, H. (2000). "Opportunities to Enhance a Commercial Web Site," Information & Management, 38(1), 15-21.

57. Weeks, P. & Crouch, I. (1999). "Sites for Sore Eyes: An Analysis of Australian Tourism and Hospitality Web Sites," Information Technology and Tourism, 2, 153-72.

58. Wei, S., Ruys, H. F., van Hoof, H. B. & Combrink, T. E. (2001). "Uses of the Internet in the Global Hotel Industry," Journal of Business Research, 54(3), 235-41.

59. Winer, R. S. (2001). Customer Relationship Management: A Framework, Research Directions and the Future, Haas School of Business, University of California, Berkeley.

60. Zafiropoulos, C., Vrana, V. & Paschaloudis, D. (2006). "The Internet Practices of Hotel Companies: An Analysis from Greece," 18(2), 156-163.

61. Zineldin, M. (2000). "Beyond Relationship Marketing: Technologicalship Marketing," Journal of Marketing Intelligence & Planning, 18, (1), 9-23.

9

USING MODIFIED IPA TO IMPROVE SERVICE QUALITY OF STANDARD HOTEL IN TAIWAN

Li-Hsing Ho[1], Shu-Yun Feng[2], Tieh-Min Yen[1*]

[1]Department of Technology Management, Chung-Hua University, HsinChu, Taiwan

[2]Departmant of Hospitality Management, Chung-Hua University, HsinChu, Taiwan

ABSTRACT

The purpose of this paper is to establish the modified IPA (Importance-Performance Analysis) methodology for service quality characteristics and improvements in the hotel industry. Traditional IPA methodology has its shortcoming as it assumes that importance-performance is a normal distribution. However, if that is not the case, the traditional IPA model cannot correctly identify the priority for improvements. Therefore, this paper runs Box-Cox Transformation to convert data into normal distributions, in order to perform statistical analyses and comparisons on a likefor-like basis. This approach avoids the potential problem of the IPA model but maintains its simple and intuitive decision-making model. This paper examines standard hotels and performs literature reviews and experts' interviews in order to establish the list of service quality characters for the hotel industry. It also conducts a survey to illustrate the application and benefits of the modified IPA method, so as to identify the key characteristics of the service quality and the ways to improve service quality of standard hotels.

KEYWORDS

Standard Hotels, Service Quality, Importance-Performance Analysis, Box-Cox Transformation, SERVQUAL

1. INTRODUCTION

Most of the studies on the constructs of service quality of hotels are based on SERVQUAL, a service quality framework developed by Parasuraman et al. [1] . Many scholars examine the constructs and characteristics of service quality of hotels and they find that there are large variances in such constructs and characteristics [2] -[9] . Some scholars establish service quality measurements exclusively for the hotel industry, e.g. LODGQUAL developed by Getty and Thompson [10] and HOLSERV developed by Wong et al. for Australian hotels [11] . However, there is no established list of constructs and characteristics of service quality for standard hotels in the context of international perspectives so that this list can serve as a benchmark for consumers, hoteliers and government agencies in the assessment of hotel performances and service quality in Taiwan.

The basic concept of IPA is to gain an understanding of the perceived importance of service quality from the customers' perspectives and experience via market surveys. The purpose is to measure the levels of satisfaction and perceived quality characteristics, establish a two-dimension analysis of importance-performance and identify the strategic actions required [12] -[14] . Since the development of marketing strategies with the use of the IPA method by Martilla and James [15] , this method has been widely applied in many industries [14] [16] -[21] . Martilla and James emphasize that the IPA model boasts a number of advantages, i.e. low costs, ease of applications, better focuses and strategic recommendations [15] . It is also possible to analyze the strengths and weaknesses of an organization based on market surveys. Therefore, this paper adopts the IPA decision-making methodology as the tool to examine the service quality of standard hotels.

Over the recent years, some scholars attempt to modify the traditional IPA model in order to make it more realistic. These innovative researches on IPA find that importance and performance are correlated [22] -[24] . Matzler et al. apply multiple regression analyses to deduct the relative importance of quality characteristics, i.e. implicit importance [25] . It is a relative importance based on coefficients.

All these scholars make significant contributions to the IPA methodology. However, the traditional IPA model has potential problems: statistical analyses and scale-based data collected usually do not meet its assumption of normal distributions [13] [16] . Garver also believes that the assumption of normal distributions is seldom realistic [26] . As a result, the use of arithmetic means of skewed data as a benchmark makes it difficult to correctly identify the priority for service quality improvements [16] . This paper proposes an IPA methodology modified with Box-Cox Transformation and illustrates its applications and benefits by examining the service quality of standard hotels.

2. LITERATURE REVIEW

2.1. Service Quality of Hotel Industry

The discussion of service quality has been based on PZB model over the recent years. Parasuraman et al. believe that service quality is measured by the gap between ex-ante expectations and ex-post perceptions [27] . They perform an exploratory study on four service industries, i.e. banking, credit cards, securities brokerage and product maintenance and establish the PZB service quality model. Parasuraman et al. conduct factory analyses to establish a service quality measurement of good credibility and validity [1] [28] . This measurement, known as SERVQUAL, consists of 5 constructs and 22 service quality questions. The five constructs are tangibility, reliability, responsiveness, assurance and empathy.

Most of the studies on the quality constructs and characteristics of the hotel industry are based on SERVQUAL developed by Parasuraman et al. [1] [28] . Akan modifies the SERVQUAL measurement with Likert 4 scales and 30 service quality constructs for the study on the hotels in Istanbul [2] . In his model, there are three service quality constructs, i.e. facilities, staff and service flows. Juwaheer examine the service quality of hotels in the perspective of international travellers with modified SERVQUAL [3] . He performs a Principal Component Analysis in order to extract 9 key characteristics from 39 service quality constructs for the hotel industry. These characteristics are reliability, assurance, perks, communication skills and values, accommodation attractiveness, considerate and accurate services, food and beverage, landscape and environments. Antony et al. examine the operational performances of hotels in the UK by adopting the PZB Gap Model and SERVQUAL to establish a list of 16 service quality characteristics [4] . They also conduct studies on service quality and operational performances from Gap 1 to Gap 4. Eccles and Durand explore the methods to improve service quality of hotels by referring Gap 5 of the PZB model and SERVQUAL and addressing practical issues [5] .

In addition to SERVQUAL and PZB model, literature reviews and experts' opinions can also derive a list of service quality characteristics. Min and Min perform a study on service quality benchmarking of the hotel industry and comes up with two service quality constructs and 14 indicators [6] . They are 1) room values: cleanness, comfort, atmosphere, toiletry, prices, space and giveaways; 2) front-desk services: manner, real-timeliness, complaint processing, reservation, reception, leisure activities and relevant guidance and services. Min et al. perform a dynamic benchmarking of service quality of the hotel industry and divide service quality characteristics into two categories with a total of 20 items [7] . The additional items for room values are quietness, desk facilities, Internet, faxing and free local calls. The additional items for front-desk services are discounts and benefits of regular customers and flexible hotel policies. Desombre and Eccles refer to the service quality studies on the hotel industry as the reference for the improvements of medical service quality [8] . They come up with 12 service quality constructs, i.e. accessibility, visibility, usability, cleanness, comfort, communication, ability, etiquette, friendliness, reliability, responsiveness and safety. Tsang and Qu perform a study on the perspectives of travellers and hoteliers regarding service quality of hotels in China and summarize 35 service quality characteristics via literature reviews and experts' opinions [9] . They are comforts and welcoming feelings, availability of service staff, professionalism, multiple languages, friendliness, etiquette, conference facilities and locations, etc. Knutson et al. develop LODGSERV, a service quality measurement for hotels [29] . Knutson et al. develop a list of 26 service quality characteristics for luxury hotels in order to measure the expectations of consumers [30] . Getty and Thompson formulate LODGQUAL, a service quality measurement for standard hotels in the US [10] . Getty and Getty come up with a list of 26 service quality characteristics in order to evaluate the quality perceived by consumers [31] . Wong et al. propose HOLSERV, a service quality measurement for hotels in Australia [11] . According to the above literature review, this paper finds large variances in service quality characteristics of the hotel industry as a whole. Due to a lack of service quality standards for hotels in Taiwan, it is hence necessary to establish a list of service quality constructs and characteristics of international perspectives, in order to facilitate the research work of this paper.

2.2. IPA Model

Martilla and James believe that market surveys can establish an understanding of the level of acceptance of specific quality characteristics by customers [15] . IPA provides a perfect solution. Data processing alone can analyze and understand four types of quality characteristics and develop strategic actions for the quality characteristics in each of the four quadrants. It is easy to interpret and formulate strategic actions accordingly, as shown in Figure 1. The interpretation of the IPA matrixes is defined with the four quadrants:

Concentrate here, C: Customers believe the product or service quality characteristic is highly important, but the organization fails to perform.

Keep up the good work, K: Customers believe the product or service quality characteristic is highly important, and the organization performs well.

Low priority, L: Customers do not think the product or service quality characteristic is important and the organization fails to perform too.

Possible overkill, P: Customers do not think the product or service quality characteristic is important, but the organization performs well.

After the publication of IPA Analysis by Martilla and James, many renowned scholars come up with modified statistical models. Yavas and Shemwell refer to the integrated relative importance as a weighted indicator to replace the importance represented with the vertical axis [32] . They also use relative performance by measuring the variance in performance of quality characteristics between the organization in question and its competitors. This is then multiplied with importance. The product is used to replace the performance represented with the vertical axis. Tarrant and Smith refer the standard deviations of the mean of quality characteristics to modify IPA [33]. Matzler and Sauerwein believe that self reports by customers cannot represent the true importance of quality characteristics [25] . They also prove that the importance as described by customers themselves is not a function of the satisfaction with quality characteristics. Explicit importance is the function of the satisfaction with quality characteristics. This is deducted from a multiple regression equation for k number of quality characteristics as performance as an independent variable (X_i) and overall satisfaction as a dependent variable (Y). The function can be expressed as $Y = \beta_0 + \beta_1 X_1 + \beta_2 X_2 + ... + \beta_k X_k + \varepsilon$, with ε denoting the error item. The coefficient β_i indicates the influence of the i-th quality characteristic on the overall satisfaction. Matzler and Sauerwein believe that the importance of quality characteristics perceived by customers should be measured with the coefficients derived with multiple regression equations [25] . Lee et al. argue that the absolute values of the coefficients should be used [19] . Meanwhile, the model fit of the regression equation is the key because it indicates the credibility of importance. Sampson and Showalter propose a dynamic model to prove the changes of organizational performances because it will also change the importance perceived by customers [24] . This is mainly because the coefficient between importance and performance is not zero.

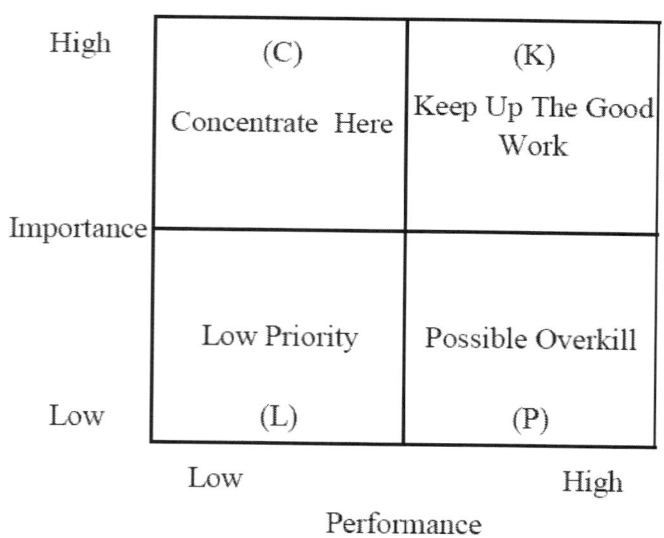

Figure 1. Importance-performance matrix.

Despite the major contributions from many scholars to the IPA methodology, the problem remains: questionnaire surveyed data do not meet with the assumption of a normal distribution [16] [18] . Therefore, this paper converts data with Box-Cox Transformation so that the converted data meet the assumption of a normal distribution. This approach modifies the traditional IPA model and avoids analysis errors and decision mistakes.

3. RESEARCH DESIGN

3.1. Research Subjects

This paper intends to construct a list of service quality constructs for standard hotels in Taiwan in order to assist them to improve service quality and satisfy the accommodation needs of customers. The research subjects include hotels, hostels, homestays, vacation centers, motels and guest houses.

3.2. Research Method

This paper sources data of the past 20 years from Emerald and Science Direct, two databases for literature analyses and generalizations. Interviews with experts are performed in order to gather opinions from consumers, hoteliers, scholars and government officials about service quality required for standard hotels. This paper summarizes a list of service quality characteristics as a basis for the questionnaire survey and issues the questionnaires on a random-sampling basis in accordance with the number of hotels in different regions. The gathered data is converted with Box-Cox Transformation to facilitate analyses in the modified IPA model. The purpose is to formulate strategies to improve service quality of standard hotels.

3.3. Questionnaire Design

The questionnaire on the importance and performance of service quality characteristics for standard hotels is measured with Likert 5 scale. The purpose is to gain an understanding of the quality perceived by customers and the performance of standard hotels. The performance is measured with the scale of 1 to 5, with 1 indicating very much dissatisfied and 5 indicating very satisfied. Similarly, the importance is also measured with the scale of 1 to 5, with 1 indicating very unimportant and 5 indicating very important. Basic data of the respondents is also required.

3.4. Box-Cox Transformation

Data conversion is critical to statistical analyses. If data is not in a normal distribution, the analysis will be biased. Data comparisons are only possible when linear relationships, independence and normality are in line. It is necessary to test data before any statistical analyses. If data are not in compliance with assumptions, data conversions become necessary [34] .

It is usually desired that converted data is close to a normal distribution and the mean and variances are independent from each other. In other words, converted data have fixed error variances. Table 1 list a number of frequently seen data conversion modes [34] :

It is necessary to understand the relationship between the means and variances of the data in order to select an appropriate data transformation mode. However, it is still difficult to perform analyses. Therefore, Box and Cox applies maximum likelihood method to deduct Equation (1) for data conversion and to estimate parameter λ [35] .

$$y^{(\lambda)} = \begin{cases} \dfrac{y^{\lambda} - 1}{\lambda \dot{y}^{\lambda-1}} & \lambda \neq 0 \\ \dot{y} \ln y & \lambda = 0 \end{cases} \tag{1}$$

Let $\dot{y} = \ln^{-1}\left[(1/n)\sum \ln y\right]$ be the geometric mean of the observed values. λ method identified the λ value that minimizes the square sum of errors, i.e. $SS_E(\lambda)$. Usually, a graph is plotted with $SS_E(\lambda)$ vs. λ to facilitate the identification of the λ value that minimizes $SS_E(\lambda)$ or deduct $L(y)$, the maximum log-likelihood function of $y_i^{(\lambda)}$. The best conversion value can maximize λ value of the maximum log-likelihood function $L(y)$, as shown in Equation (2).

Table 1. Commonly seen data conversion methods.

Relationship of σ_y and μ	β	$\lambda = 1 - \beta$	Conversion
$\sigma_y \propto \; cons\tan t$	0	1	No transformation
$\sigma_y \propto \mu^{\frac{1}{2}}$	$\frac{1}{2}$	$\frac{1}{2}$	Square root
$\sigma_y \propto \mu$	1	0	Log
$\sigma_y \propto \mu^{\frac{3}{2}}$	$\frac{3}{2}$	$\frac{-1}{2}$	Reciprocal square root
$\sigma_y \propto \mu^2$	2	−1	Reciprocal

$$L(y) = -\frac{1}{2} n \ln\left[\frac{1}{n}\sum_{i=1}^{n}\left(y_i^{(\lambda)} - y^{(\bar{\lambda})}\right)^2\right] + (\lambda - 1)\sum_{i=1}^{n}\ln(y_i) \tag{2}$$

Let $y^{(\bar{\lambda})} = \frac{\sum_{i=1}^{n} y_i^{\lambda}}{n}$

Box-Cox Transformation is known for its compliance of the basis assumption for linearility, normality and homomoscedasticity. It is also a widely applied method [36] . This paper converts the data collected with IPA questionnaires with Equation (1) and Equation (2) on important-performance, so that the data become compliant with the statistical assumptions.

3.5. IPA Model Modified with Box-Cox Transformation

Hansen and Bush suggest that IPA is considered a simple and effective tool [37] . It can help to decide how to effectively utilize limited resources and achieve the maximum customer satisfaction. This paper converts the two-axis IPA dimensions into statistics of normal distributions, as shown inFigure 2. Central tendency is measured with means, and means are referred to as the cut-off values for the two axes forming four quadrants. The definitions of the four quadrants are the same as the IPA proposed by Martilla and James [15] . This makes interpretation and decision-making easy and straightforward.

This paper recovers the questionnaires and converts the data with Equations (1) and (2), in order to establish a modified IPA matrix for analyses and comparisons. The purpose is to provide the information required for service quality improvements and action plans in the hotel industry.

4. RESULTS & ANALYSIS

4.1. Service Quality Characteristics of Standard Hotels

This paper refers to the five service quality constructs developed by Parasuraman et al. [1] [28] and develops the list of service quality characteristics by sourcing data of the past 20 years from Emerald and Science Direct. Table 2 shows the list of service quality characteristics established with literature reviews and interviews with 12 experts, hoteliers, scholars and government officials [7] [9] [31] [38] [39] .

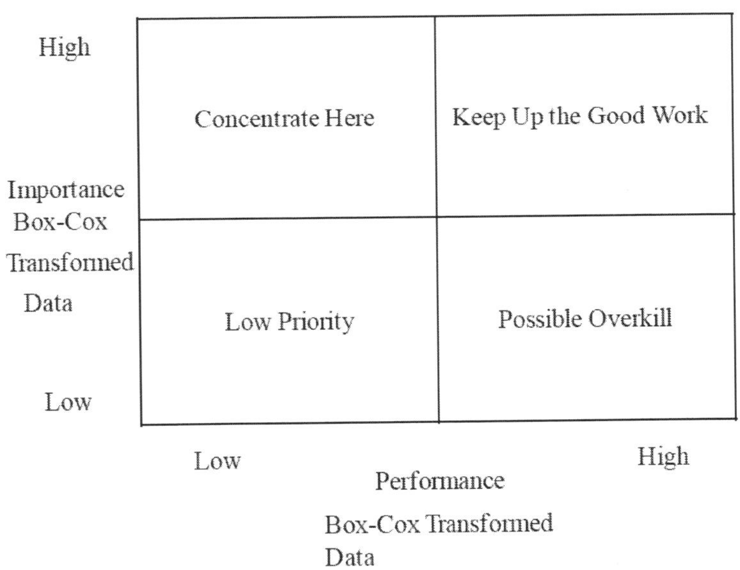

Figure 2. IPA matrix modified with Box-Cox Transformation.

Most of the interviewed hoteliers, scholars and government officials indicate that the list of service quality characteristics summarized on the basis of literature reviews can serve as a starting point for service quality assessments. However, some hoteliers, scholars and government officials suggest that "hotel marketing" and "merchandize selling" should be added to the construct of "tangibility". Hotel marketing refers to advertising and websites. Merchandize selling refers to the sale of souvenirs. Hoteliers interviewed suggest that "appropriate support to staff to provide better services" should be added to the construct of "assurance". Scholars and government officials recommend that "deployment of barrier-free space and toilets for mother and children" should be added to the construct of "empathy". Table 2 summarizes the list of service quality characteristics for the questionnaire survey based on the interviews with experts.

4.2. Reliability & Validity Analysis

This paper performed random sampling of standard hotels in different regions by issuing 350 questionnaires in Taiwan. With 317 questionnaires recovered, the recovery rate was 90.571%. After the elimination of 39 invalid questionnaires, this paper has 278 valid questionnaires.

This paper performs analyses on reliability and validity after the recovery of the questionnaires. The Cronbach's α values of respective importance constructs are 0.768 for tangibility, 0.757 for reliability, 0.955 for responsiveness, 0.931 for assurance and 0.882 for empathy. The Cronbach's α for importance, 0.920, is higher than 0.7, and hence it is considered to be of high reliability [40] . This shows that the research content has a high degree of internal consistency. The factor loading of respective constructs is 0.448 for tangibility, 0.351 for reliability, 0.849 for responsiveness, 0.876 for assurance and 0.384 for empathy. The values for reliability and empathy are close to 0.4, and the values for the other three constructs are all above 0.4. The P-value of the reliability construct is 0.062, but those of other constructs are all 0.000 < 0.001, reaching statistical significance. This shows that these variables carry convergent validity.

Similarly, the Cronbach's α values of respective satisfaction constructs are 0.897 for tangibility, 0.842 for reliability, 0.909 for responsiveness, 0.918 for assurance and 0.860 for empathy. The Cronbach's α for satisfaction, 0.950, is higher than 0.7, and hence it is considered to be of high reliability [40] . This shows that the research content has a high degree of internal consistency. The factor loading of respective satisfaction constructs is 0.651 for tangibility, 0.751 for reliability, 0.816 for responsiveness, 0.817 for assurance and 0.824 for empathy. All the values are above 0.4 and P-values are all 0.000 < 0.001, reaching statistical significance. This shows that these variables carry convergent validity.

4.3. IPA Analysis Post Box-Cox Transformation

Take room comfort and atmosphere perceived by customers (T1) for example, the Kolmogorov-Smirnov test on importance and performance results in P-values of 0.118 > 0.05 and 0.010 < 0.05, respectively. Apparently, the importance of room comfort and atmosphere (T1) meets the assumption of a normal distribution, but the performance does not. This paper calculates the means of importance and performance of service quality perceived by customers according to the traditional IPA method and the results are summarized in column 3 (importance) and column 4 (performance) in Table3 As Table 2 shows, the mean values of importance and performance of room comfort and atmosphere (T1) perceived by customers are 4.270 and 3.766, respectively. The mean values of importance and performance of quality characteristics are 3.995 and 3.658, respectively. The numbers are referred to as the basis for IPA quadrants. The service quality characteristics are divided into four categories. Take room comfort and atmosphere (T1) for example, this quality characteristic falls in the quadrant of high importance (4.270 > 3.995) and high performance (3.766 > 3.658). This means hotels should keep up the good work (K Strategy) for this service quality. Column 5 of **Table 2** shows the strategic actions for other service quality characteristics.

Figure 3 illustrates a traditional IPA matrix, with most of the quality characteristics falling into K (Keep up the good work) and L (Low priority). Few of them are in C (Concentrate here) or P (Possible overkill) and all close to the intersection at the central point. It is difficult to determine, in Tarrant and Smith [33] , what the strategic actions should be called for. This paper refers to the average performance of quality characteristics as the independent variable and the average importance of quality characteristics as the dependent variable for the regression analysis. Adjusted R2 is 0.747. The F value in the ANOVA is 110.334, reaching statistical significance. This proves that importance and performance are correlated [22] -[24] , and hence it is difficult to determine strategic actions in a traditional IPA matrix, as argued by Sampson and Showalter [24] .

Table 2 . Service quality characteristics of standard hotels.

Construct	Assessment
	Room comfort and atmosphere (T1)
	Unique décor (T2)
	Good views and landscape (T3)
	Neatness (T4)
Tangibility	Geographic location (T5)
	Parking space, size and convenience (T6)
	Low prices (T7)
	Well-groomed personnel (T8)
	Modern facilities (Karaoke, TV, newspapers and magazines, T9)
	Comprehensive facilities (swimming pools, air-conditioners, conference rooms, saunas and Internet, T10)
	Hotel marketing (advertising, websites, etc., T11)
Tangibility	Merchandize selling (souvenirs and specialty, T12)
	Quiet rooms (T13)
	Safety (REL1)
	Hotel publicity (REL2)
Reliability	Room privacy (REL3)
	Timely completion of promises made to customers (REL4)
	Friendly staff (RES1)
	Willingness to assist customers (RES2)

Responsiveness	Ability to resolve problems for customers in a timely manner (RES3)
	Services meeting the expectations of customers (RES4)
Assurance	Ability of service staff to provide correct information (A1)
	Professionalism of service staff (A2)
	Trustworthiness of service staff (A3)
	Customers feeling secure when interacting with service staff (A4)
	Appropriate support to staff to provide better services (A5)
	Mutual assistance among service staff to facilitate better services (A6)
Empathy	Simple procedures to make reservations (E1)
	Convenient accommodation (E2)
	Convenient transportation (E3)
	Consultation in sightseeing (E4)
	Deployment of barrier-free space and toilets for mother and children (E4)
	Food & beverage (E5)
	Whether employees speak foreign languages (E6)
	Personalized attention from service staff (E7)
	Prioritization of customers' best interest (E8)
	Service hours addressing the needs of customers (E9)
	Understanding of special needs of customers (E10)

This paper converts the data with Box-Cox Transformation into 5 scales and normal distributions, in order to modify the IPA model. Take the importance of room comfort and atmosphere (T1) for example, this paper calculates the best conversion value with Equation (2) that allows the λ to be at the maximum of 1.965 in maximum log-likelihood function of $L(y)$. This value introduced in Equation (1) derives the Box-Cox Transformation value of 2.720. This paper calculates the mean of the importance of room comfort and atmosphere (T1) perceived by customers to be 3.200. The means of the perceived importance of other service quality characteristics are calculated in the same way and the results are shown in Column 6 of Table3 Similarly, the performance of room comfort and atmosphere (T1) is calculated with Equation (2). The best conversion value allows the λ to be at the maximum of 2.052 in maximum log-likelihood function of $L(y)$. This value introduced in Equation (1) derives the Box-Cox Transformation value of 1.905. This paper calculates the mean of the performance of room comfort and atmosphere (T1) perceived by customers to be 1.765. The means of the perceived performance of other service quality characteristics are calculated in the same way and the results are shown in Column 7 of Table3

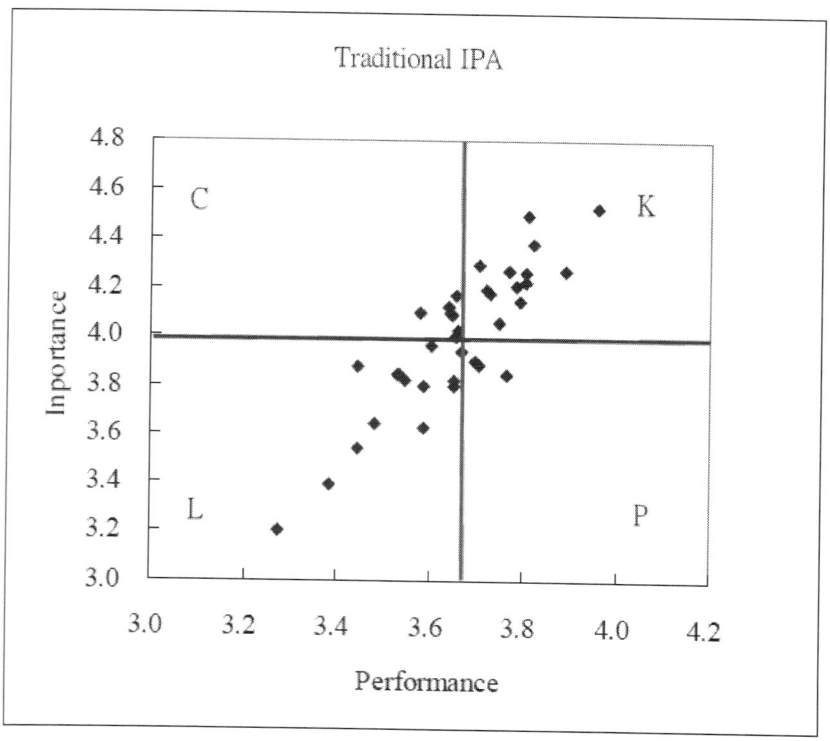

Figure 3. Traditional IPA matrix.

Table 3. IPA and BC-IPA analysis results.

Code	Dimensions of Service Quality	Imp.	Per.	IPA Category	BC-Imp	BC-Per	BC-IPA Strategy
T1	Room comfort and atmosphere	4.270	3.766	K	3.200	1.765	L
T2	Unique décor	3.644	3.482	L	2.948	2.029	L
T3	Good views and landscape	3.820	3.651	L	2.469	2.063	L
T4	Neatness	4.529	3.957	K	11.929	2.318	C
T5	Geographic locations	3.799	3.651	L	3.232	3.799	P
T6	Parking space, size and convenience	3.799	3.586	L	2.239	3.281	L
T7	Low prices	3.960	3.601	L	2.654	3.774	P
T8	Well-groomed personnel	3.845	3.763	P	2.459	4.701	P
T9	Modern facilities (Karaoke, TV, newspapers and magazines, etc.)	3.626	3.586	L	2.490	3.351	L
T10	Comprehensive facilities (swimming pools, air-conditioners, conference rooms, saunas and Internet, etc.)	3.824	3.547	L	3.114	3.063	L
T11	Hotel marketing (advertising, websites, etc.)	3.392	3.385	L	2.931	2.538	L
T12	Merchandize selling (souvenirs and specialty)	3.201	3.273	L	8.896	1.688	C
T13	Quiet rooms	4.270	3.888	K	3.348	5.648	P
REL1	Safety	4.493	3.809	K	11.039	5.545	K
REL2	Hotel publicity	3.543	3.442	L	3.721	3.478	C
REL3	Room privacy	4.381	3.820	K	4.806	5.261	K
REL4	Timely completion of promises made to customers	4.295	3.701	K	3.535	5.323	K

RES1	Friendly staff	4.201	3.784	K	2.951	5.450	P
RES2	Willingness to assist customers	4.176	3.727	K	2.834	4.559	P
RES3	Ability to resolve problems for customers in a timely manner	4.086	3.647	C	2.580	4.203	P
RES4	Services meeting the expectations of customers	4.119	3.640	C	2.691	4.215	P
A1	Ability of service staff to provide correct information	4.000	3.655	C	2.358	3.671	L
A2	Professionalism of service staff	4.000	3.655	C	2.342	3.600	L
A3	Trustworthiness of service staff	4.025	3.658	K	2.404	3.685	L
A4	Customers feeling secure when interacting with service staff	4.054	3.748	K	2.483	4.515	P
A5	Appropriate support to staff to provide better services	3.881	3.705	P	2.302	3.754	P
A6	Mutual assistance among service staff to facilitate better services	3.899	3.694	P	2.249	3.141	L
E1	Simple procedures to make reservations	4.223	3.806	K	3.056	4.609	P
E2	Convenient accommodation	4.259	3.806	K	3.245	5.401	P
E3	Convenient transportation	4.140	3.791	K	2.980	4.203	P
E4	Consultation in sightseeing	3.842	3.532	L	2.393	3.636	L
E5	Deployment of barrier-free space and toilets for mother and children	3.874	3.446	L	2.492	3.256	L
E6	Food & beverages	4.094	3.579	C	2.683	3.784	P

E7	Whether employees speak foreign languages	3.845	3.529	L	2.639	3.113	L
E8	Personalized attention from service staff	3.942	3.665	P	2.557	1.606	L
E9	Prioritization of customers' best interest	4.165	3.655	C	2.892	3.584	L
E10	Service hours addressing the needs of customers	4.191	3.719	K	2.984	4.673	P
E11	Understanding of special needs of customers	4.097	3.644	C	2.728	3.269	L
Mean		3.995	3.658		3.444	3.725	

According to Figure 4 and Table 3, the analysis based on the modified IPA model shows that safety (REL1), room privacy (REL3), and timely completion of promises made to customers (REL4) are in the quadrant of high importance and high performance. This means hotels should keep up the good work (K) in order to maintain competitive advantages. The service quality characteristics falling into the quadrant of high importance but low performance are neatness (T4), merchandize selling (souvenirs and specialty, T12), and hotel publicity (REL2). This means should concentrate her (C) immediately. The service quality characteristics falling into the quadrant of low importance and low performance are room comfort and atmosphere (T1), unique décor (T2), good views and landscape (T3), Parking space, size and convenience (T6), Modern facilities (Karaoke, TV, newspapers and magazines, T9), Comprehensive facilities (swimming pools, air-conditioners, conference rooms, saunas and Internet, T10), hotel marketing (advertising, websites, etc., T11), Ability of service staff to provide correct information (A1), professionalism of service staff (A2), trustworthiness of service staff (A3), consultation in sightseeing (E4), Deployment of barrier-free space and toilets for mother and children (E5), Whether employees speak foreign languages (E7), personalized attention to customers (E8), prioritization of customers' best interest (E9) and understanding of special needs of customers (E11). These items are low priorities (L) that do not have high business risks. The service quality characteristics falling into the quadrant of low importance but high performance are geographic locations (T5), low prices (T7), well-groomed staff (T8), quiet rooms (T13), friendly staff (RES1), willingness to assist customers (RES2), ability to resolve problems for customers in a timely manner (RES3), services meeting the expectations of customers (RES4), customers feeling secure when interacting with service staff (A4), appropriate support to staff to provide better services (A5), mutual assistance among service staff to facilitate better services (A6), simple procedures to make reservations (E1), convenient accommodation (E2), convenient transportation (E3), food & beverages (E6), and service hours addressing the needs of customers (E10). These items are possible overkills (P). There is no need to invest any additional resources, but efforts should be made to other areas such as C and K [14] [18] .

4.4. Discussion

This paper refers to Tarrant and Smith [33] for the consideration of standard errors of the means of service quality characteristics, and focuses its discussion on the service quality characteristics falling into specific quadrants based on the data converted with Box-Cox Transformation. Neatness (T4) falls into K (Keep up the good work) in the traditional IPA analysis. However, the IPA analysis modified with Box-Cox Transformation finds it in the quadrant of C (Concentrate here). According to a study by Lewis and McCann [41] on failed services of hotels, business and leisure travellers think dirty rooms are the worst problem of all. Min et al. suggest that room neatness is the most important factor perceived by customers [7] . Tsang and Qu argue that there is a gap, statistically significant, between their expectations and the perception of room neatness [9] . Therefore, hoteliers must focus on this quality in order to avoid complaints from customers. Yang et al. examine business and leisure hotels in

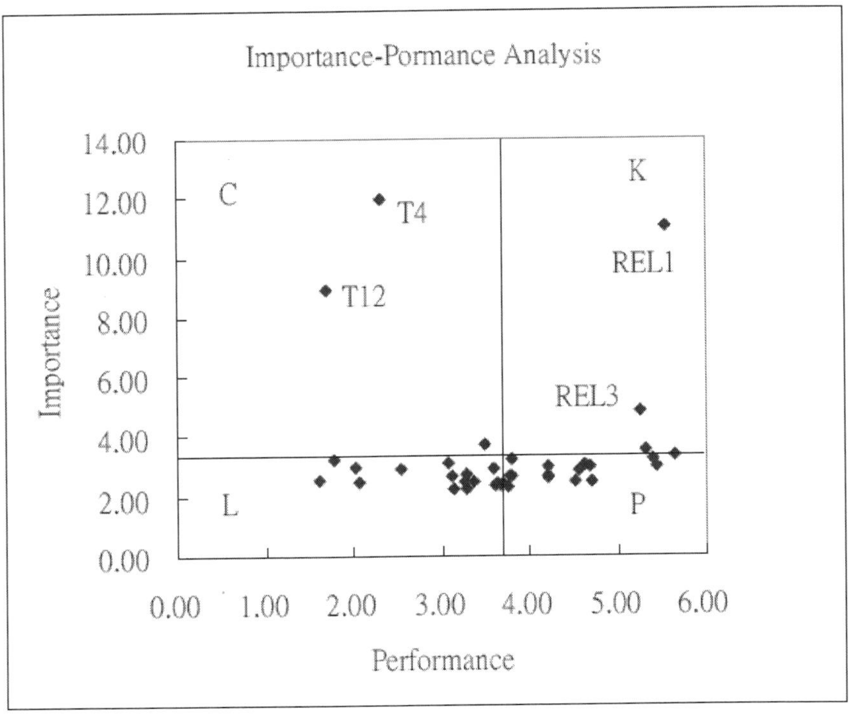

Figure 4. IPA matrix converted with Box-Cox Transformation.

Taiwan and find that neatness is a one-dimensional quality [38] . The better the neatness (T4), the higher customers' satisfaction is. The traditional IPA decision-making model will determine it is to keep the good work. However, the IPA analysis with data converted with Box-Cox Transformation prompts hoteliers to continue to invest resources to improve neatness (T4), in order to develop differentiation and create competitive advantages [14] [42] [43] .

Merchandize selling (souvenirs and specialty, T12) falls into L (Low priority) in the traditional IPA analysis. However, the IPA analysis based on data converted with Box-Cox Transformation finds it fall into C (Concentrate here). This paper refers to the literature on service quality of hotels during the past 20 years and does not find the establishment of merchandize selling (souvenirs and specialty) as a service item. Meanwhile, according to the interviewed experts, most of the hotels do not sell merchandize (souvenirs and specialty, T12). Ottenbacher and Harrington believe that the servicing process is one of the innovations for hotels [44] . Therefore, they should sell merchandize (souvenirs and specialty, T12). The changes in customers' perceptions can also be a source of service quality innovations. It is possible to improve performance of particular services and develop differentiation to create competitive advantages [14] . IPA analysis based on the data converted with Box-Cox Transformation highlights the importance of merchandize selling (souvenirs and specialty, T12).

Safety (REL1) and room privacy (REL3) fall into K (Keep up the good work) under both the traditional IPA analysis and the modified IPA analysis based on the data converted with Box-Cox Transformation. Tsang and Qu examine the hotel industry in China and find that the gap between the expectation and perception of safety (REL1) is small [9] . This means Chinese hotels perform well on this score. According to a study by Lewis and McCannon failed services of hotels, dangerous facilities and environments do not rank at the top of the complaint list of business and leisure travellers [41] . This means this quality meets the expectations of customers. Tseng examine the tourism hotels in Taiwan and suggest that safe environments and room privacy are causal criteria that influence the changes of other factors and serve as a competitiveness benchmark [39] . Yang et al. investigate the business and leisure hotels in Taiwan and find that safety (REL1) is a one-dimensional factor [38] . The safer it is, the higher the customers' satisfaction. It is important and valuable to customers so hotels should keep up the good work to maintain competitive advantages[13] [45] .

5. CONCLUSIONS

This paper established 38 service quality characteristics for standard hotels after literature reviews and interviews with industry representatives, government officials, scholars and experts. Questionnaires were issued to consumers, and the service quality of the hotels was analyzed with the IPA model modified with Box-Cox Transformation. This paper concludes the following: It is necessary to "keep up the good work" (K) in safety (REL1) and privacy (REL3) of accommodation, in order to maintain competitive advantages. It is necessary to "concentrate here" (C) in neatness (T4) and sale of merchandize (souvenirs) (T212), in order to create differentiated market segments or the source of innovation.

This paper makes the following contributions: 1) Establishment of the list of constructs and service quality characteristics for standard hotels, as a reference for consumers, hoteliers and government agencies in terms of performance assessments; 2) The IPA method modified with Box-Cox Transformation to make data in compliance with the assumption for a normal distribution. This approach lays out the foundation for the IPA method in statistical analyses.

This paper establishes a measurement for service quality of standard hotels via literature reviews and experts' interviews. Perspective of consumers is referred to in order to ensure the robustness of this list. It is suggested that future studies can perform factory analyses for exploratory analyses and empirical studies, in order to establish measurements of service quality for standard hotels. The IPA methodology modified with Box-Cox Transformation enables data to be in compliance with the assumption for a normal distribution. It is possible to apply the data converted with Box-Cox Transformation in the IPA linear model proposed by Matzler and Sauerwein [25] , Kano Model proposed by Lee et al. [16] and Gap Analysis suggested by Parasuraman et al. [1] [28] . Meanwhile, this paper only focuses on the constructs and assessment items of service quality for standard hotels in Taiwan. It is recommended that future studies can expand the scope of service quality research to the hotel industry as whole, in order to enhance the service quality of hotels in Taiwan.

REFERENCES

1. Parasuraman, A., Zeithaml, V.A. and Berry, L.L. (1988) SERVQUAL: A Multi-Item Scale of Measuring Consumer Perceptions of Service Quality. Journal of Retailing, 64, 12-41.

2. Akan, P. (1995) Dimensions of Service Quality: A Study in Istanbul. Managing Service Quality, 5, 39-43.

3. Juwaheer, T.D. (2004) Exploring International Tourist's Perceptions of Hotel Operations by Using a Modified SERVQUAL Approach—A Case Study of Mauritius. Managing Service Quality, 14, 350-364.

4. Antony, J., Antony, F.J. and Ghosh, S. (2004) Evaluating Service Quality in a UK Hotel Chain: A Case Study. International Journal of Contemporary Hospitality Management, 16, 380-384.

5. Eccles, G. and Durand, P. (1997) Improving Service Quality: Lessons and Practice from the Hotel Sector. Managing Service Quality, 7, 224-226.

6. Min, H. and Min, H. (1997) Benchmarking the Quality of Hotel Services: Managerial Perspectives. International Journal of Quality & Reliability Management, 14, 582-597.

7. Min, H., Min, H. and Chung, K. (2002) Dynamic Benchmarking of Hotel Service Quality. Journal of Service Marketing, 16, 302-321.

8. Desombre, T. and Eccles, G. (1998) Improving Service Quality in NHS Trust Hospitals: Lessons from the Hotel Sector. International Journal of Health Care Quality, 11, 21-26.

9. Tsang, N. and Qu, H. (2000) Service Quality in China's Hotel Industry: A Perspective from Tourist and Hotel Managers. International Journal of Contemporary Hospitality Management, 12, 316-326.

10. Getty, J.M. and Thomson, K.N., (1994) A Procedure for Scaling Perceptions of Lodging Quality. Hospitality Research Journal, 18, 75-96.

11. Wong, A.O.M., Dean, A.M. and White, C.J. (1999) Analyzing Service Quality in the Hospitality Industry. Measuring Service Quality, 9, 136-143.

12. Hu, H.Y., Chiu, S.I., Cheng, C.C. and Yen, T.M. (2011) Applying the IPA and DEMATEL Models to Improve the Order-Winner Criteria: A Case Study of Taiwan's Network Communication Equipment Manufacturing Industry. Expert Systems with Applications, 38, 9674-9683.

13. Hu, H.Y., Lee, Y.C. and Yen, T.M. (2009) Amend Importance-Performance Analysis Method with Kano's Model and DEMATEL. Journal of Applied Sciences, 9, 1833-1846.

14. Lee, Y.C., Cheng, C.C. and Yen, T.M. (2008) Integrate Kano's Model and IPA to Improve Order-Winner Criteria: A Study of Computer Industry. Journal of Applied Sciences, 9, 38-48.

15. Martilla, J.A. and James, J.C. (1977) Importance-Performance Analysis. Journal of Marketing, 41, 77-79.

16. Lee, Y.C., Yen, T.M. and Tsai, C.H. (2008) Modified IPA for Quality Improvement: Taguchi's Signal-to-Noise Ratio Approach. The TQM Journal, 20, 488-501.

17. Huang, Y.C., Wu, C.H. and Hsu, C.J. (2006) Using Importance-Performance Analysis in Evaluating Taiwan Medium and Long Distance National Highway Passenger Transportation Service Quality. Journal of American Academy of Business, 8, 98-104.

18. Lee, Y.C., Hu, H.Y., Yen, T.M. and Tsai, C.H. (2008) Kano's Model and Decision Making Trial and Evaluation Laboratory Applied to Order-Winners and Qualifiers Improvement: A Study of Computer Industry. Information Technology Journal, 7, 702-714.

19. Lee, Y.C., Yen, T.M. and Tsai, C.H. (2008) Using Importance-Performance Analysis and Decision Making Trial and Evaluation Laboratory to Enhance Order-Winner Criteria: A Study of Computer Industry. Information Technology Journal, 7, 396-408.

20. Tonge, J. and Moore, S.A. (2007) Importance-Satisfaction Analysis for Marine-Park Hinterlands: A Western Australian Case Study. Tourism Management, 28, 768-776.

21. Lee, Y.C., Yen, T.M. and Tsai, C.H. (2008) The Study of an Integrated Rating System for Supplier Quality Performance in the Semiconductor Industry. Journal of Applied Sciences, 8, 453-461.

22. Oh, H. (2001) Revisiting Importance-Performance Analysis. Tourism Management, 22, 617-627.

23. Ryan, C. and Huyton, J. (2002) Tourists and Aboriginal People. Annals of Tourism Research, 29, 631-647.

24. Sampson, S.E. and Showalter, M.J. (1999) The Performance-Importance Response Function: Observations and Implications. The Service Industry Journal, 19, 1-25.

25. Matzler, K. and Sauerwein, E. (2002) The Factor Structure of Customer Satisfaction: An Empirical Test of Importance Grid and the Penalty-Reward-Contrast Analysis. International Journal of Service Industry Management, 13, 321-332.

26. Garver, M.S. (2002) Using Data Mining for Customer Satisfaction Research. Marketing Research, 14, 8-12.

27. Parasuraman, A., Zethaml, V.A. and Berry, L.L. (1985) A Conceptual Model of Service Quality and Its implications for Future Research. Journal of Marketing, 49, 41-50.

28. Parasuraman, A., Zeithaml, V. and Berry, L.L. (1991) Refinement and Reassessment of the SERVQUAL Scale. Journal of Retailing, 67, 420-450.

29. Knutson, B., Stevens, P., Wullaert, C., Patton, M. and Yokoyama, F. (1991) Lodgserv a Service Quality Index for the Lodging Industry. Hospitality Research Journal, 14, 277-284.

30. Knutson, B., Stevens, P., Patton, M. and Thompson, C. (1993) Consumers' Expectation for Service Quality in Economy, Mid-Price

and Luxury Hotels. Journal of Hospitality and Leisure Marketing, 1, 27-43.

31. Getty, J.M. and Getty, R.L. (2003) Lodging Quality Index (LQI): Asseeeing Customers' Perceptions of Quality Delivery. International Journal of Contemporary Hospitality Management, 15, 94-104.

32. Yavas, U. and Shemwell, D.J. (2001) Modified Importance-Performance Analysis: An Application to Hospitals. International Journal of Health Care Quality Assurance, 14, 104-110.

33. Tarrant, M.A. and Smith, E.K. (2002) The Use of a Modified Importance-Performance Framework to Examine Visitor Satisfaction with Attributes of Outdoor Recreation Settings. Managing Leisure, 7, 69-82.

34. Montgomery, D.C. (2007) Introduction to Statistical Quality Control. Wiley India Pvt. Ltd., New Delhi.

35. Box, G.E.P. and Cox, D.R. (1964) An Analysis of Transformation. Journal of the Royal Statistical Society, Series B (Methodological), 26, 211-252.

36. Sakia, R.M. (1992) The Box-Cox Transformation Technique: A Review. Journal of the Royal Statistical Society, Series D (The Statistician), 41, 169-178.

37. Hansen, E. and Bush, R.J. (1999) Understanding Customer Quality Requirements: Model and Application. Industrial Marketing Management, 28, 119-130.

38. Yang, C.C., Jou, Y.T. and Cheng, L.Y. (2009) Using Integrated Quality Assessment for Hotel Service Quality. Quality and Quantity, 24, 349-364.

39. Tseng, M.L. (2009) Using the Extension of DEMATEL to Integrate Hotel Service Quality Perceptions into a Cause-Effect Model in Uncertainty. Expert Systems with Applications, 36, 9015-9023.

40. Nunnally, J.C. (1967) Psychometric Methods. McGraw-Hill Book Company, New York.

41. Lewis, B.R. and McCann, P. (2004) Service Failure and Recovery: Evidence from the Hotel Industry. International Journal of Contemporary Hospitality Management, 16, 6-17.

42. Kano, N., Seraku, N., Takanashi, F. and Tsjui, S. (1984) Attractive Quality and Must-be Quality. The Journal of the Japanese Society for Quality Control, 14, 39-48.

43. Matzler, K., Hinterhuber, H.H., Bailon, F. and Sauerwein, E. (1996) How to Delight your Customers. Journal of Product Brand Management, 5, 6-18.

44. Ottenbacher, M.C. and Harrington, R.J. (2010) Strategies for Achieving Success for Innovative Versus Incremental New Service. Journal of Services Marketing, 24, 3-15.

45. Hu, H.Y., Lee, Y.C., Yen, T.M. and Tsai, C.H. (2009) Using BPNN and DEMATEL to Modify Importance-Performance Analysis Model: A Study of Computer Industry. Expert Systems with Applications, 36, 9969-9979.

10

MARKETING STRATEGIES FOR HOTEL INDUSTRY INTERNATIONALIZATION IN MORELIA

Oscar H. Pedraza Rendón, Rubén Molina Martínez, María S. Ramírez Flores*

Universidad Michoacana de San Nicolás de Hidalgo, Instituto de Investigaciones Económicas y Empresariales

ABSTRACT

The marketing in this industry has gained big protagonist in today's context. Therefore, there are many studies that have started to analyze both the causes and the forms in which this process takes place and how the industry benefits from it. Internationalization views from the perspective of the Mexican economy could be a benefit; sustaining and increasing the amount of exterior market, improving the internal and external competitive level and in the end, guaranteeing the possibility of economy, market and employment (Garcia, J.M. 1996), this is why it is important to analyze how marketing can trigger that internationalization. The processes of industrial internationalization are one of the most solid ways to develop competitive advantages, promoting international industrial competitivity (Canals, J. 1992) and strengthening the acquired position in the domestic market. Maybe because it is a new process, the area that covers the internationalization of the hotel industry in Morelia has not been the object of scientific and monographic analysis from the marketing point of view. The objective is to propose strategies of marketing that promote the internationalization of hotel industry in Morelia. Analyses how the touristic activities were impacted by marketing, also includes a revision of the most relevant studies that, from a theoretical or empirical point of view of the strategies and the internationalization, making emphasis on the relationship of the two phenomena. The theme that is object of study focuses its analysis in the academic and industrial currents. The results of the field work that was carried out for the study applied on the process of internationalization of the hotel industry in Morelia; have to be shown, as well as the strategies of marketing that the hotel industry must apply to obtain the so called internationalization

KEYWORDS

Marketing strategies, internationalization, hotel industry

1. TOURISM AND HOTEL INDUSTRY

Traditionally, the hospitality companies and the small business dedicated to the hospitality have been apart of the marketing environment. However, factors such as crisis, hyper-segmentation and digitization in all types of promotion, are causing the emergence of multiple cases of successful business models developed by them. In all, the marketing treated as strategy and brand value creation play a role.

This is why social unite economic situation demands that the enterprise and the revitalization of the hotel are the engine to return to revive the economy. To do this, the hotel industry should moreliana consider marketing as the focus of its business integrator.

The World Tourism Organization confirms the good situation in which tourism is worldwide. Expanding the number of countries consider tourism as a potential source of profit and therefore the stimulus to develop this activity has increased both the number and the attraction of new destinations. If one also includes the national market entry of international players (Kim, C.; HWang, P. (992), competition intensifies it more difficult to establish a strategy to achieve a sustainable positioning companies within this complex business environment (Garcia, J.M. 1996).

This has joined demand more demanding and informed and with very different profiles between segments (Varela, F.J.; Thompson, E.; Y Rosch, E. 2004). Having driven the tourism industry to carry out differentiation strategies through the use of marketing, because it is one of the main factors influencing the competitiveness of the tourism sector (Blankenburg, D.; Johanson, J. 1997). The hotel industry is important for its contribution to job creation and the indirect effects it has on the environment. The presence of a hotel in a certain area can promote the development of complementary tourist activities, improve the welfare of residents in the tourist areas of influence, and revitalize a number of economic activities, without the figure of the hotel could disappear.

The hotels in Morelia, Michoacan, are essentially family businesses that are developing their administrative skills, they face competition from chain hotels, which have a low participation of foreign tourists, providing a hosting service that has failed standard retain guests, and lack of training and management systems of information useful for decision-making, so it is relevant to increase occupancy standards applying new strategies to ensure the continuous flow of foreign customers.

2. THEORETICAL PERSPECTIVES ON INTERNATIONALIZATION TO BUSINESS

Internationalization is an economic phenomenon that has attracted the interest of a large number of researchers. For internationalization means a set of operations that facilitate the establishment of more or less stable ties between the company and international markets along a process of increasing involvement and internationally (Welch, L; Luostarinen, R. 1988).

There are different theoretical approaches that attempt to explain the process of internationalization of companies, those that refer to the explanations relating to the analysis of international operations of the multinational company, and more specifically, its direct investment activities.

Table 1. Definition of business internationalization and foreign market selection for each approach described

INTERNATIONALIZATION OF THE COMPANY	APPROACHES THEORETICAL	DECISION OF SELECTION FOREIGN MARKETS
Based on the concentration processes of companies across national borders due to market structures (Wilson, D.T., Y Moller, K.E.K. 1991).	Tª of the organization industrial	Production facilities are located in selected foreign markets in terms of the competitive advantages and the company has exclusive
It relies on overseas markets to internalize be more efficient because it avoids the costs associated with external market imperfections	paradigm eclectic Dunning	It's a rational decision based on costs and in the advantages of producing in certain markets
It explains how foreign production pattern of multinational companies and to explain why it makes foreign direct investment	Approach macroeconomic	It is based on the search of countries with comparative advantages in sectors or activities which make direct investment
It relies on overseas markets to internalize be more efficient because it avoids the costs associated with external market imperfections	Tª of the internalization	It is based on the identification of the location advantages in foreign markets in terms of economic factors: costs, economies of scale, market structure
It is a gradual process of cumulative decisions based on permanent interactions between the development of knowledge about foreign markets and the commitment of resources in those markets	Model Uppsala	Determines which countries should address the company based on perceived psychological distance
It is a business innovation process based on the export performance of the Company	Approach innovation	It is based on the analysis of more favorable external conditions for export activities
Stage sampling is a process of new product development conditioned by the life cycle of products	Life cycle model of Vernon	It decides where to locate production in terms of economic conditions evaluated: economies of scale, cost minimization, per capita income, demand, enterprise assets
It is a process that develops in terms of continuous interorganizational interactions between local firms and their networks	Network approach	It is based on the interaction between policy makers and members of social networks to reduce search costs overseas markets

Source: Authors.

2.1 Internationalisation from a network approach

The growing importance of the dynamics of international trade relations is demonstrated in the emerging number of studies based on the approach of network theory (Ellis, P. 2000). Much of the literature to date has focused on the development of relationships (Canals, J. 1994), while the beginning of these relationships has not been studied usual (Olsen, M. 1991). Following extended network theory, entry into foreign markets is seen as a function of continuous interorganizational interactions between local companies and their networks. This means that the opportunities of foreign markets will reach the local company by members of the network (Blankenburg, D.; Johanson, J. 1997).

The authors believe that the first mover has a low level of knowledge about foreign market conditions and also can not use their relationships with other members of the network to acquire, as they have not been internationalized, its activities develop in a national or local network (Olsen, M.; Connolly, D. 2000). This implies that the company will assume a certain risk that seek to minimize using a penetration strategy, which is to cover the nearby markets through input methods that involve a low level of commitment. However, companies can support initiating various strategies to get better results getting the necessary experience through marketing.

Marketing has many definitions, according to Philip Kotler is "social and managerial process by which groups of individuals meet their needs by creating and exchanging goods and services."

3. ANALYSIS OF THE HOTEL SECTOR IN MORELIA, FROM THE PERSPECTIVE OF MARKETING

A mistake made by many hotels, when they have begun to be aware of the need to structure the marketing function has been to concentrate on the operational side (advertising, design, catalogs, etc.), and have neglected the strategic approach customer orientation and brand value creation (branding strategy) (Olsen, M.; West, J.; Tse, E. 1998).

In the SME customer focus has been in the hands of commercial nature of the entrepreneur who led the project. This guidance has been deposited in your intuition and business sense. His leadership has supplied strategic focus, structure and method.

On the branding strategy, in most cases is limited to the design process of the formal elements of the brand. They have neglected phases needed branding process as defining identity platform, the development of the value proposition and the definition of the desired positioning (Olsen, M.; Zhao, J.L. 1997).

3.1 Features Tourism Demand Michoacan

Data from a survey conducted in 2010 by the National Tourism Secretariat to identify the profile of tourists visiting the state of Michoacan.

The following summarizes the main features:

- These are young tourists. 48% of visitors were aged between 25 and 44 years.
- Most of the demand comes from the domestic market. The proportion of Mexican tourists reached 85% of the total.
- Of the tourists from abroad, 63% originated in the United States of America.
- The most form commonly used is the journey with a partner (40%).
- The tourists stayed overnight an average of 4.55 nights in Michoacan. This is a key indicator for tourism and in particular for hosting service providers.
- The transport most used to reach Michoacán was the private car (77%).
- Among the reasons given by the tourists to visit Michoacan, the reason "cultural events" obtained a significant increase in recent years and now accounts for 40% of the total.

ACTIVITY IN TURISM CENTERS SELECTED BY CATEGORY 2011 p/

	Rooms		%	Arrival of Tourists	Tourists Night
	Available	Occupied	Occupation	Total	Total
Total National	**128,739,035**	**64,402,167**	**50.03**	**59,237,051**	**133,359,849**
5 Estrellas	51,750,193	30,267,773	58.49	19,903,501	64,221,274
4 Estrellas	33,381,333	17,291,692	51.80	16,841,535	34,842,868
3 Estrellas	23,051,083	9,245,150	40.11	11,377,361	19,418,063
2 Estrellas	11,111,031	4,306,332	38.76	5,994,641	8,590,615
1 Estrella	9,445,395	3,291,220	34.84	5,120,013	6,287,029
Morelia, Mich.	**1,431,175**	**577,245**	**40.33**	**705,336**	**980,259**
5 Estrellas	257,300	106,846	41.53	93,932	164,866
4 Estrellas	593,976	288,572	48.58	322,795	482,099
3 Estrellas	335,377	101,542	30.28	163,113	180,096
2 Estrellas	148,130	53,981	36.44	85,204	103,066
1 Estrella	96,392	26,304	27.29	40,292	50,132

SOURCE: Office of Tourism of the State governments and the Federal District.

3.2. Keys to Marketing Strategy

1. Commitment from senior management Analysing the success stories of large companies and brands today are the leaders, in most cases is the figure of some business leaders who were able to turn their vision, their dream a reality business. The leader's role is crucial in a hotel regardless of the size it is. Now when this is starting or when it is still small, is when the role of the highest level of management is most important. Morelia Hotels rarely have this commitment to address because they are smaller hotels run by individual owners and not qualified to run a business (Root, F.R. 1994).

2. First strategy, then operational. In practice, it is very common to find companies where the marketing department called endeavors to produce catalogs and brochures, preparing attendance at fairs, feed web page content and little else. This mentality is wrong and pernicious. The first should be the strategic focus and that we must focus early efforts. A proper analysis of the market, customers, competition, internal and context, enabling accurate segmentation to segment with the greatest chance of success, differential positioning can be transmitted with the resources available and that results in a defensible advantage, a selection successful channel with a pricing policy which rewards the channel and allow sustainable profitability, excellence in service embodying a winning proposition customer value. It should foster strategic culture in the hotel industry and to achieve what until today

seems impossible, plan, not only the internal activities of the hotel, but everything that can achieve better business performance.

3. Customer orientation. The "customer orientation" is one of the common in the literature on management and marketing. Generally, there is a consensus that the hotel industry should be "customer oriented". Morelia is known for its hospitality and quality of service, so if being planned or be part of a strategy, small family hotels if Estonian linked with such guidance.

4. Digitization. This is the factor that historically represents the greatest opportunity for hotels to compete with weapons similar to those of large enterprises. While it is true that in recent years has gone deep into the internet, we are far from digitization rates of countries like United States, England and Germany. And of course, this fact is an obstacle to the competitiveness of SMEs.

So much so, that there have been initiatives like NEW (No Company without Web) and similar to encourage these companies to start in the digital economy.

Not enough to have their own web presence or website of the third, but to make a deep reflection about the opportunities that the Internet can offer the business and what is the right strategy to take advantage. Some of the benefits are:

- Success can be fast.
- They break geographical barriers.
- Segmentation varies.
- The function of advertising, promotion and public relations changes.
- Internet marketing allows real results. • The size is still important
- Customers.
- Competition.

The reality is that 70% of family hotels do not have a website and much less efficient a digitized system for tourist attraction, so it is necessary to manage training and moderenización in this area.

5. Innovation. Marketing as a process of creating value from the company or brand to its public. This value contribution may arise from a stagnant or from one approach innovative. Hotel size, agility and responsiveness, along with the possibilities of the digital economy, must be aligned by management in order to build a culture of innovation. Innovation must be then maximum marketing in SMEs and entrepreneurs. Today it has placed greater emphasis on innovation, Morelia seeks to be competitive for the variety of services and their quality in terms of accommodation, so that daily seeks continuous innovation in the service.

6. Talent and collaboration. In such a complex environment, digitized, competitive and demanding innovation, you can only compete and create value for customers (ie marketing), if the organization has the necessary talent. For hotels, its size will be impossible in most cases available talent internally and know-how necessary to understand, interpret and act on such a complex scenario. It is therefore necessary to establish a culture of collaboration. It is not worth a company closed in on itself, jealous of their knowledge and information not shared, not with your customers and suppliers, but not even with their employees. No easy task, as it involves a cultural change, direct responsibility senior management, which is based on the following keys (Olsen, M.D. 1993):

- Confidence
- Information for key people.
- Innovation
- Generosity
- Values

4. PROPOSAL FOR MARKETING IN THE HOSPITALITY INDUSTRY

That the current situation is a complicated and reality uncertain, both large and small companies. It is also a fact that this is the environment in which it will have to compete for a long time. Gone are the initial phase of retrenchment in any crisis is to review structures appraise costs and weather the storm as best as possible.

Current demand to take action. This means concentrating on income, that is, on the client. To do this, hotels and hoteliers should adopt, once and for all, marketing and brand building as the focus of its activities. This means focusing efforts on offering to its clients and prospects, innovative value propositions, differential and winning.

The circumstances of undoubted opportunities to exploit these opportunities should bet on the marketing from senior management. Must have strong leadership that from a strategic, bet on guiding the company to the customer, both domestic and abroad, although this involves giving up short-term profits.

Also, this strategic leadership must establish a culture of innovation and proceed to reflection about digitization. All this will lead to a series of requirements of talent, experience and human and financial resources that the hotel is hard to tell, therefore, this leadership should promote the establishment of collaborative environments and strategic alliances with third parties, so that to ensure that their value proposition is supported by a back-office talent.

The hotels in the city of Morelia given its status as small or family enterprises, should consider the six main keys to marketing to internationalize your hosting service. As established in the network approach to internationalization, you should get companies that have services or products for hotels, rejan collaborative networks and implement marketing together to achieve increased market share and meet international.

4.1 Proposal

4.1.1 Price Strategies

One of the strategies that must be used by the hotels is the price for seasonality . This means that prices should be set in line with the season of the year when tourists visit the country. The objective of this strategy should always be to increase income in each season encouraging tourists visit in any season.

The parity pricing strategy is also recommended. This strategy implies that the prices offered are similar to those of the competition (in this case other countries that offer tour packages). It is convenient to apply because the tourism product offered is similar to that of competitors, but there is an advantage in the state that can be leveraged to provide the consumer with the highest overall value: the chance to visit forests, beaches, ruins archaeological sites, picturesque villages, all within walking distance.

Promotion program is prepared based on a process in which various aspects are contemplated. This process consists of five steps.

Step1: Define the promotional objectives. Create a strong identity of the place as a tourist destination in the consumer market and develop the investment potential in the market and sale of the hotel industry, strongly encouraging its members to take an active interest in developing and selling the country's tourism programs.

Step 2: Promotional Strategy Selection. Promotional strategies adopted are:

a) Attraction Strategy: This strategy is ideal for capturing directly to consumers, which is one of the promotional plan objectives. Normally this strategy is characterized by having in its promotional mix , with a greater emphasis on advertising activities.

b) Push Strategy: This strategy is used because you want to stimulate product sales strongly tourism , encouraging the actors associated with the item to be interested in promoting it. For the case, the mixture Coupon is characterized by a higher content of activities relating to sales promotion.

Step 3: Determination of the promotional mix. Promotional mix using includes:

1) Direct Mail.

2) Making promotional visits.

3) Conducting familiarization trips.

4) Running an advertising campaign in the industry.

5) Participation in events of the Tourism Industry.

Step 4: Preparation of individual programs. Each of the actions listed in the previous section, require the preparation of individual programs, which will be implemented.

Step 5: Budget Media. A media budget depends on the promotional schedule, which includes all promotional activities to be performed, with specific dates

4.1.2 Selective Distribution Strategy.

Mainly tourism operators hotels and travel agencies should seize national and international trips to trade fairs to be known and establish contact with foreign operators to achieve select to distribute packages to Morelia.

5. CONCLUSIONS

Faced with this situation, que highlights the growing intensification of competition That is taking place in the industry, reflecting Both investment fever that is living in the hotel industry in Mexico,: such as the landing of large international chains in this Country, it that required the large number of small and medium hotel companies Consider new marketing strategies to Improve their competitiveness and enable to continue playing the important role in this industry have developed in recent medical support.

In this sense, one of the lines of action to follow should consider joining the so-called vertical marketing systems, in which without losing their independence are integrated into distribution groups allow a better competitive that when negotiating position with large distribution groups, higher profitability and efficiency in marketing activities, and ultimately from Improve their competitive position.

Similarly, you must pass what is known as a political marketing targeted intensive distribution strategy where instead of working with a very large number of intermediaries is chosen by choosing a smaller number of them, but thereby maintaining a high level of cooperation in the distribution, establishing long-term partnerships and agreements Between the companies involved.

Third, a vital strategy for small and medium enterprises through the goal of customer loyalty, que is required to develop a clear strategy of differentiation from the competition, constantly looking for a way to offer something different from the other and overall consumer satisfaction. In this sense, an aspect of great Importance is to enhance the brand image, allowing a clear positioning of the company to the consumer and constant innovation policy consumers that look positively surprise each time you stay at the hotel.

REFERENCES

1. Blankenburg, D.; Johanson, J. (1997): "Business network connections and the atmosphere of international business relationships", en BJÖRKMAN, I.; FORSGREN, M., (1997) op.ct.
2. Canals, J. (1992): L'internacionalització, Departament d'Industria, Generalitat Canalunya, Barcelona.
3. Canals, J. (1994): La internacionalización de la empresa. Cómo evaluar la penetración en mercados exteriores, McGraw-Hill, Madrid.
4. Ellis, P. (2000), "Social Ties and Foreign Market Entry", Journal of International Bussines Studies, Vol. 31 (3), pp. 443-469.
5. Garcia, J.M. (1996), Dirección estratégica: Fundamentos, CIES, Caja Insular de Ahorros de Canarias.
6. Johanson, J., Y Vahlne, J.-E. (1993), "Management of internationalization", en Zan, L.; Zambon, S.; y Pettigrew, A.M. (Eds.), Perspective on Strategic Change, Kluwer Academic Publishers, Boston, pp. 42-78.
7. Kim, C.; HWang, P. (1992): "Global Strategy and multinationals entry mode choice", Journal of International Business Studies, vol. 23, nº1, pp.29-53.
8. Kim, K Y Olsen, M. (1999): "Determinants of successful acquisition processes in the US lodging industry", International Journal of Hospitality

9. Kim, W.C.; Arbel A. (1998): "Predicting merger targets of hospitality firms (a Logit model)", International Journal of Hospitality Management, nº17, pp.303-318.

10. Olsen, M. (1991): "Structural Changes: The international hospitality industry and firm", International Journal of Contemporary Hospitality Management. Vol. 3, nº 4, pp 21-24, MCB University Press.

11. Olsen, M.; Connolly, D. (2000): "Experience-based travel: how technology is changing the hospitality industry", Cornell hotel and Restaurant Administration Quarterly, February, pp.30-40, Cornell University.

12. Olsen, M.; West, J.; Tse, E. (1998): "Strategic Management in the hospitality industry, second edition, Jonh Wiley & Sons, New York.

13. Olsen, M.; Zhao, J.L. (1997): "New management practice in the international hotel industry", EIU Travel and Tourism Analyst, nº 1, pp.53- 73. Olsen, M.D. (1993): "International Growth strategies of major US hotel companies.", E.I.U. Travel and Tourism Analyst, nº3, pp.51-64.

14. Root, F.R. (1994), Entry Strategies for International Markets, Lexington Books, Nueva York.

15. Varela, F.J.; Thompson, E.; Y Rosch, E. (2004), The Embodied Mind: Cognitive Science and Human Experience, MIT Press.

16. Welch, L; Luostarinen, R. (1988): "Internationalization: evolution of a concept", Journal of General Management, vol. 14, pp.34-55.

17. Wilson, D.T., Y Moller, K.E.K. (1991), "Buyer-seller relationships: Alternative conceptualisations", en Paliwoda, Stanley J (Editor), New Perspectives on International Marketing, Routledge, Londres, pp. 87-107.

11

INNOVATIVE FOOD SAFETY STRATEGIES IN A PIONEERING HOTEL

Su-Ling Wu

Department of Leisure Management, Yu Da University, Miaoli County, Chinese Taipei.

ABSTRACT

A case study approach was adopted to identify the innovative food safety strategies in place at one pioneering hotel that had voluntarily implemented food safety control. An investigation of food safety strategies and the reasons for their implementation in the hotel foodservice were carried out using multiple sources of data, including interviews with key decision makers in the hotel, observations of the business environment, and a review of documentation. The findings suggest that not only food control strategies but also marketing and corporate strategies are important when addressing the problems of food safety. The findings of this study also demonstrate the relationship between motivating factors and food safety strategies. Analysis of the interviews indicates that a hotel's food safety strategies depend significantly on the attitude of senior management, the firm's capability, and corporate image.

KEYWORDS

Food Safety Strategies; Food Safety Control System (FSCS); Hotel

1. INTRODUCTION

Despite the advances in technology and hygiene standards in the past three decades, the outbreaks of fatal food born illness continued to increase globally, such as outbreaks of Salmonella serotype Enteritidis (SE), Listeria monocytogenes (Lm) and Bovine Spongiform Encephalopathy (BSE) in Europe and Northern America; Escherichia coli O157:H7 in Japan; cholera in Asia, Africa, and Latin America [1]. Life style changes and rapid global economic growth contributed to this trend [1]. These two factors lead to great numbers of people eating processed or imported food, and consuming food in foodservice establishments; and the increase of global movement in food supply and people also increase the exposure rate of contacting food born illness due to the unawareness of food born hazard on imported food or on food consumption while traveling aboard [1]. To help curb the rate of food born illness generated by the food industry, government agencies in Taiwan and other countries continue to pass new regulations, such as mandatory or voluntary enforcements of Good Manufacturing Practices (GMP), Good Hygienic Practices (GHP), Hazard Analysis Critical Control Points (HACCP) and ISO 9000 and 22000 series. Moreover, other stakeholders, such as food safety advocacy groups, customers, the press and media, and others, are pushing for firms to take food safety strategies from farm-to-fork, such as risk assessments throughout the external supply chains and the internal production process proactively into their managerial decision making process to minimize contamination which lead to loss of reputation, revenue, and even disruption of business. The incidence of food borne illness may damage the reputation of a country as a tourist destination [2]. The dining experiences of tourists were among the most important predictors of overall travel satisfaction [3,4]. Other researches indicated that the perception of food risks at their travel destinations also influenced their purchasing decisions for other products, especially when they were traveling to less developed countries [5,6]. Therefore, implementation of a certified food safety standard such as the GHP and HACCP standards would benefit especially less developed countries or regions, like Taiwan, by offering food quality assurance, strengthening the foodservices' competitiveness and ultimately enhance the country or region's reputation as tourist destination. In Taiwan approximately 40% of food borne illness outbreaks was linked to commercial food establishments (http://www. fda.gov.tw). Due to these repeated outbreaks of food borne illness, the Taiwanese Health Department in 2000 published a Food Safety Control System (FSCS), which includes principles from both HACCP and Good Hygienic Practice (GHP), in order to bring Taiwan in line with international standards. Beginning in

September 2008, the Taiwanese Heath Department has been promoting the FSCS standards with stringent guidelines for the hotel industry. These standards have been gradually taking effect in outlets with central kitchens or kitchens in international tourist hotels. In order to promote the Taiwanese tourism industry, since 2009 the Taiwan Tourism Bureau has provided financial incentives to encourage the hotel industry to improve their facilities and service quality to meet international standards by obtaining local or internationally recognized certificates such as FSCS (http://admin.taiwan.net.tw).

Consumers react strongly to negative reports of food safety and their purchasing behaviors are negatively influenced [7]. Consumer food safety concerns have increased rapidly in recent years, as several high profile food incidents grabbed the head lines. Nevertheless serious incidents continue to occur. Food safety concerns have become an important factor for consumers in determining food selection and consumption. Chemical food additives, meats from diseased pigs and pesticide residuals are considered to be the most severe food safety problems by consumer in Taiwan, especially while they dine in commercial food outlets and consume processed food or meat products [8]. Similar results were also found in two studies which indicated that food safety issues influenced consumers' dining-out decisions [9,10]. Fears of being featured on the media and press were one of reasons for foodservice initiating food safety control program [11]. Therefore, the foodservice industry should try to build their reputation by offering less pre-packaged, processed food and more on healthy, nutritious fresh certified produces. Food Safety Control System (FSCS) specifies the requirements needed for food safety management to ensure the final food is safe for consumption. Therefore, the innovative food safety strategies currently adopted and the strategies being carried out in a pioneering hotel are the focus of this study.

Previous hospitality food safety research has focused only on the food safety control strategy. However, that the levels and types of food safety control strategy include not only a functional strategy (control and marketing strategy), but also a business (competitive advantage) and a corporate strategy (a link with quality and other corporate objectives) [12,13]. The main aim of this study was to understand the food safety strategies (activities) in a hotel food service, while government bureaus have been gradually putting requirements and standards in place, and encouraging tourist hotels to gain FSCS certification. This study explores experiences of the implementation and operation of FSCS in a pioneering international tourist hotel in Taiwan, and thus identifies the key motivating factors and the challenges to implementing food safety strategies.

2. METHOD

An investigation of food safety strategies and the reasons food safety strategies are or are not implemented in hotel food services was carried out using multiple sources of data, including interviews, observation of environmental conditions, and the review of documentation to increase the construct validity and reliability of this study [14]. The method was designed to yield results that would assist the prediction of future drivers of food safety strategies and their uptake in the hotel food service industry. A franchised international tourist hotel that had voluntarily implemented the Food Safety Control System (FSCS) was identified as a suitable participant for the purposes of the study.

Semi-structured interviews were conducted with senior management in the hotel, such as managers, chefs, and food safety supervisors, who were in management teams that had received special training in food safety control system principles [13]. The interviewer (the researcher) specifically directed open-ended questions to the interviewee in relation to food safety strategies. A set of questions was as follows:

1) What are the food safety practices in your company? How are they carried out in your firm? Why?

2) Is it convenient for you to practice food safety strategies? If not, how could it be made convenient for you?

3) What is your biggest food safety management difficulty?

4) Is there anything that would assist you in your food safety management efforts?

With the consent of the participants, interviews were taped and later transcribed. A pattern matching analytical procedure was used to increase the internal validity of a study [14]. It was anticipated that this analysis would lead to a set of food safety strategies being identified and their motivators.

The case study was conducted in a hotel that was a locally owned franchise of an international hotel chain. The chain's first hotel opened in the early 1990s in Taiwan and there are now 9 hotels in the chain. The company has been striving to reinforce food safety strategies voluntary since early 1999 in all its hotels. It is company policy that all hotels in the chain adopt a food safety control system (FSCS). The hotel that participated in this study obtained FSCS certification in 2011.

3. RESULTS AND DISCUSSION

Seven interviews with staff from the chain hotel were conducted for the case study. The interviewees included managers in the food and beverage departments, executive chefs, food safety supervisors, personnel managers and general managers. Most had been in the hotel industry for more than 10 years. All of them had received training in food safety and received FSCS certification. The average time per interview was 35 minutes. Observations of environmental conditions and examination of documentation such as annual reports, policy statements, employee manuals, product advertisements, and corporate advertisements were also collected in order to obtain a greater understanding of operational systems. Many interesting issues were found during analysis of the interviews. The interview data revealed that the study hotel engaged in many food safety strategies, such as safety control, marketing, and corporate organization.

3.1. Safety Control Strategies

The study hotel's capacity to provide sufficient personnel, a training program, time, and resources was found to be the main barrier to the consistent use of a food safety control system [15,16]. As far as possible, however, the principles of the FSCS were adhered to by the various departments of the hotel, as well as the hotel's suppliers. Production equipment; cleaning and sanitation; personal hygiene; training; chemical control; receiving, storage, and shipping; traceability and recall; and pest control (http://food.fda.gov.tw) were all subject to provisions outlined in the FSCS. The interviewees indicated that the hotel's Food Safety Control System (FSCS) could be successfully carried out with proper layout and equipment and an effective training and assessment program.

3.1.1. Layout and Equipment

Analysis of the data from the interviews revealed that proper layout and equipment in the kitchen promoted food safety and supported food safety systems. With a more centralized kitchen, the interviewees indicated that less cost, better food quality, and less food borne risk could be achieved [13].

"Our hotel has a centralized kitchen in the basement of the building where food ingredients are pre-processed and supplied to all the kitchens in the hotel. Cost can be reduced through easier control in inventory, better deal in purchasing with large quantities, and less labor and equipment needed in each kitchen. Food quality is easier to control, not only in food freshness, texture, and portion size, but also in food safety. Different areas for different kinds of food in process and storage are designed to eliminate cross contamination. Examples of physical separation include the space of food preparation and refrigeration separated according to the food type such as seafood, meat, cold process food, hot process food, and bakery goods." (Food and Beverage Manager, 20/07/2010, interview at the hotel office).

3.1.2. In-House Training and Assessment

Effective food safety training should be designed to closely match employees' expectations and the organization's goals and be accompanied by inducements that motivate staff in order to optimize training performance [13,17,18].

"In Taiwan, kitchen staffs need to have eight hours food safety seminar attendance each year or 32 hours in four years to keep chef certification valid. FSCS certified kitchen needs to have 80% of kitchen staffs with chef certifications. In addition, at least three staffs need to have additional 32 hours FSCS training course and pass qualify exam at the end of the course in order to form a FSCS team." (Executive chef, 20/07/2010, interview at the hotel office).

For the food safety seminar, overall, the interviewees agreed that effectiveness of training is an important factor in gaining the employees' commitment to the company's goals in terms of the FSCS. However, all the interviewees agreed that the current training courses and certification awarded by the HACCP associations should offer a balance of practical skills development and knowledge requirements [13].

Proper food safety handling skill should also be included in training courses and assessed by both theory and a practical exam. Overall, the interviewees indicated that hotel staff who attended the HACCP courses all felt that the information sessions were overlong long and that the content was boring.

"The idea that the HACCP course focuses on both GHP and HACCP plan development seems ok theoretically. However, it is hard for our staffs to sit and listen eight hours a day. An interactive short course should be developed by incorporating more case discussion after showing real examples of HACCP operation in pictures or video rather than too much written words theory. In addition, practical skill development on food preparation should also be included." (Personnel Manager, 20/07/2010, interview at the hotel office).

An effective training system should not be without effective food safety management and compliance assessment. The interviewees indicated that hotel food safety was supervised and audited by an assistant manager in the Stewarding department who worked closely with the executive chef, all restaurants and kitchens, and was fully empowered by the General Manager to communicate across departments on the issue of food safety. The General Manager of the hotel was very conscious of the food safety issue, and had considerable experience in the operation of hotel kitchens and FSCS implementation. He always checked and made corrections on the food safety operation while walking around the hotel, and often walked with the food safety supervisor. A video recording system was also used to assist the assessment of activities in the hotel.

"Our hotel has video camera around the food process area, and the food safety auditor will review the video to see any defaults during operation. The auditor will also walk around the kitchen floor, take pictures, and make correction. The audit report and video will be discussed with the kitchen staff during weekly meeting in which the staff will be able to identify the problem from practical examples. It is also hope to increase the staff involvement through the weekly discussion. The practical examples can also be a good material for food safety training." (Food safety auditor, 20/07/2010, interview at the hotel office).

The Area Manager from the franchisor's head office also conducted spot-checks of food safety standards implementation. In addition, every restaurant in the hotel formed an audit team, which then assessed the other restaurants on a monthly rotation, and reported to the executive chef. The Executive Chef in an interview indicated that temperature measuring and recording were especially important for FSCS implementation, but that not all the chefs were able to do the measuring and recording work. However, the senior management ensured that they received assistance.

"Many experienced chefs, especially in Chinese cuisine, usually have lower levels of education. Some of them have difficulty to read and write and do not have hospitality relevant degrees. Therefore, now we have kitchen stewards to assist the chef on measuring and record keeping to overcome this problem." (Executive chef, 20/07/2010, interview at the hotel office).

As for environmental issues, many interviewees indicated that wearing disposable gloves was not necessary because, unlike the circumstances in a factory, most of their products were ready to eat after a short processing time. They noted that the bacteria on hands after proper washing was not enough to cause sickness, while the disposable gloves not only caused environmental problems, but also allowed employees to forget to actually wash their hands.

3.2. Marketing Strategies

Many studies [9,10,19,20] have shown that awareness of food safety or having the experience of food poisoning did influence consumers' dine out decisions.

3.2.1. Transparent/Open Kitchen

The interviewees indicated that the customers' trust concerning food safety and quality maintained or improved the company's brand image [13]. The presentation of transparent and open kitchens staffed with employees well trained in food safety was a sign to customers that food was being prepared in a safe way and that there was nothing to hide.

3.2.2. Advertising

The hotel advertised their food safety efforts by posting where their food came from and when they received FSCS certification on both their hotel's webpage and restaurant entrances.

3.2.3. In-House Training for the Public

In one interview, a General Manager indicated that the hotel held in-house FSCS training courses sponsored by one of the HACCP associations, and the training was not limited to their hotel staff, but that others could participate.

"At the end of the course, the trainee will take turn to observe and assess our Hotel FSCS system, then form a discussion after that. It is another way to advertising our hotel by having this activity." (General Manager, 20/07/ 2010, interview at the hotel office).

The interviewee indicated that they hoped the customers and the trainee could judge the positive nature of the food safety effort from the dining environment and staff presentation.

3.3. Corporate Strategies

3.3.1. Corporate Social Responsibility

In this study addressing food safety issues was found to be part of the company's policy and was documented on the mission and vision statements of the company. All the chain hotels were required to obtain food safety control system (FSCS) certification. The interviewees agreed that protecting the health and safety of the customers and employees by providing a healthy and safe environment was part of the hotel's corporate social responsibility (CSR). Many food safety efforts made by the company are outlined in annual reports and highlighted on the hotel webpage.

3.3.2. Management Team

A FSCS management team was led by the hotel General Manager and consisted of the executive committees and other heads of department, including the Food and Beverage Manager, Chief Engineer Controller, Executive Chef, Sales and Marketing Manager, Personnel Manager, Restaurant manager, Banquet Manager, Stewards, Bar Manager. The FSCS team met monthly to discuss data gathered during assessments of the hotel, and areas needing further improvement. The General Manager indicated that FSCS was a task that needed all departments to work together in order to effectively achieve required standards of food safety.

3.4. Food Safety Strategies and Their Motivators

Analysis of the interviews indicates that the commitment of senior management to food safety, the firm's capacity to act, and ideas of competitive advantage have the most positive effect on overall food safety strategies. **Table 1** shows extracts of significant food safety strategies and primary motivators from the interview data.

The commitment of the management of a hotel to standards of food safety is the key driving factor for making the resources available for the implementation of a food safety control system [13]. In Asia, most corporations still operate on a traditional hierarchical basis in which decisions are made from the top [21]; therefore, commitment on the part of senior management is especially important. The implementation of a food safety control system in the hotel was mainly influenced by the senior management, who realized that FSCS needed the cooperation of all departments. The interviewees also indicated that a firm's willingness to support FSCS relied on the proper provision of facilities and an in-house food safety training program with incentives to retain and motivate employees were significant motivators in promoting food safety program.

Table 1. Thematic conceptual matrix-food safety strategies and motivator.

	Food safety Strategies	Motivators
Safety Control Strategies	Layout & equipment In-house training/assessment	Senior management, firm's capability, Senior management, firm's capability
Marketing Strategies	Transparent/open kitchen Advertisement In-house training for public	Corp image, product differentiation, Corp image, Corp image
Corporate Strategies	CSR management team	Corp image, Senior management

The objectives of competitive advantage are cost reduction and differentiation [22], such as developing sound safe food products to improve reputation and brand loyalty. As mentioned earlier, cost saving can be easily achieved, as well as food quality and safety, with a centralized kitchen. However, interviewees indicated that the overall cost saving was minimal, compared to the investment cost. The interviewees stated that the senior management all realized the international trend toward food safety management and were willing to invest.

The franchising company's policy was to have all the hotels in the chain obtain food safety control system (FSCS) certification in order to create a brand image which customers could trust in terms of food safety and quality. Although the system was mostly carried out behind the centralized kitchen, transparent and open kitchen facilities in the dining areas were built for customers to see the hotel's product differentiation and efforts to achieve food safety and quality. Advertisement on the hotel webpage and in-house training for the public on food safety issues were also used to improve the hotel's image. One interviewee indicated that it was senior management's intention to place food safety issues in the company's strategic report, which can maintain or improve a company's image. It was also the company's policy to have a management team led by top management to run the food safety strategy more efficiently.

4. CONCLUSIONS

The main aim of this study was to understand the food safety strategies (activities) in a pioneering hotel, and the associated relationships between the motivating antecedents and food safety strategies and to determine the key motivating factors in order to provide a source of new hypotheses for further study.

This study has provided details on "what" and "how" with respect to carrying out the food safety strategies and on "how" top management commends each motivating factor. The effect of perceptions of corporate image, a firm's capability, and senior management's commitment to a food safety strategy in a large international chain hotel provided insight into the issues faced when attempting to encourage or mandate stronger safe food control systems. The results of this study suggest the most effective food safety practices and information for helping the hotel industry to further improve food safety strategies or practices. These findings can also be employed as a basis for further review by policy makers and educational institutions to ensure that the policy is appropriate for the hotel industry.

This study identified three key motivating factors— corporate image, firm's capability, and senior management's commitment—which can be developed as new hypotheses for further testing and refinement. This reduction process of data by identifying key factors is significant for forming future research.

Moreover, the result found in this study representing the relationship between the motivating factors and food safety strategies can be applied to other sectors of the food service industry or other countries confronted by similar situations and problems. The intention of this study was not to draw conclusions from the limited sample, but to provide direction for future large scale research. Therefore, any findings from the analysis of the interviewees' responses have not been used to generalise.

REFERENCES

1. World Health Organization, "International Travel and Health," 2002. http://www.who.int/ith/ITH2002.pdf

2. World Health Organization, "Food Safety," 1999. http://www.who.int/foodsafety/publications/general/brochure_1999/en/

3. C. G. Q. Chi and H. Qu, "Examining the Relationship between Tourists' Attribute Satisfaction and Overall Satisfaction," Journal of Hospitality Marketing & Management, Vol. 18, No. 1, 2009, pp. 4-25.

4. B. Ozdemir, A. Aksu, R. Ehtiyar, B. Çizel, R. B. Çizel and E. T. İçigen, "Relationships among Tourist Profile, Satisfaction and Destination Loyalty: Examining Empirical Evidences in Antalya Region of Turkey," Journal of Hospitality Marketing & Management, Vol. 21, No. 5, 2012, pp. 506-540.

5. G. Fuchs and A. Reichel, "Tourist Destination Risk Perception: The Case of Israel," Journal of Hospitality & Leisure Marketing, Vol. 14, No. 2, 2006, pp. 83-108.

6. Y. Yilmaz, Y. Yilmaz, E. T. İçigen, Y. Ekin and B. D. Utku, "Destination Image: A Comparative Study on Pre and Post Trip Image Variations," Journal of Hospitality Marketing & Management, Vol. 18, No. 5, 2009, pp. 461-479.

7. E. G. McKeown and W. B. Werner, "Content Analysis of Consumer Confidence in Food Service in Relation to Food Safety Laws, Publicity, and Sales," Journal of Hospitality Marketing & Management, Vol. 19, No. 1, 2009, pp. 72-81.

8. M. F. Chen, "Consumer Trust in Food Safety—A Multidisciplinary Approach and Empirical Evidence from Taiwan," Risk Analysis, Vol. 6, 2008, pp. 1553-1569.

9. Radam, M. L. Abu and M. R. Yacub, "Consumers' Perceptions and Attitudes towards Safety Beef Consumption," The IUP Journal of Marketing Management, Vol. IX, No. 4, 2010, pp. 29-55.

10. Knight, M. R. Worosz and E. C. D. Todd, "Dining for Safety Consumer Perceptions of Food Safety and Eating Out," Journal of Hospitality & Tourism Research, Vol. 33, No. 4, 2009, pp. 471-486.

11. J. Sneed and D. Henroid, "HACCP Implementation in School Foodservice: Perspectives of Foodservice Directors," The Journal of Child Nutrition & Management, 2003. http://docs.schoolnutrition.org/ newsroom/jcnm/03spring/sneed/

12. S. B. Banerjee, "Corporate Environmental Strategies and Actions," Management Decision, Vol. 39, No. 2, 2001, pp. 36-44.

13. S. L. Wu, "Factors Influencing the Implementation of Food Safety Control Systems in Taiwanese International Tourist Hotels," Food Control, Vol. 28, No. 2, 2012, pp. 265-272.

14. R. K. Yin, "Case Study Research: Design and Methods," 2nd Edition, Sage, Thousand Oaks, 1999.

15. P. J. Panisello and P. C. Quantict, "Technical Barriers to Hazard Analysis Critical Point," Food Control, Vol. 12, 2001, pp. 165-173.

16. L. A. Brannon, V. K. York, K. R. Robert, C. W. Shanklin and A. D. Howells, "Appreciation of Food Safety Practices Based on Level of Experience," Journal of Foodservice Business Research, Vol. 12, No. 2, 2009, pp. 1-32.

17. G. C. Ogbeide, "A Case Study of Restaurant Training Motivations and Outcomes," International Journal of Tourism and Hospitality Research, Vol. 19, No. 1, 2008, pp. 72-177.

18. J. Salazar, H. R. Ashraf, M. Tchebg and J. Antun, "Food Service Employee Satisfaction and Motivation and the Relationship with Learning Food Safety," Journal of Culinary Science & Technology, Vol. 4, No. 2, 2005, pp. 93-108.

19. Knight, E. C. D. Todd and M. R. Worosz, "Dining for Safety Consumer Perceptions of Food Safety and Eating Out," Journal of Hospitality & Tourism Research, Vol. 33, No. 4, 2009, pp. 471-486.

20. U. Z. A. Ungku Fatimaha, H. C. Boo, M. Sambasivan and R. Salleh, "Foodservice Hygiene Factors—The Consumer Perspective," International Journal of Hospitality Management, Vol. 30, No. 1, 2011, pp. 38-45.

21. Y. Boshy, "Action Learning Worldwide," Palgrave, New York, 2002.

22. R. Welford, "Corporate Environmental Management 1: Systems and Strategies," Earthscan, London, 1998.

12

THE RELATIONSHIP BETWEEN TURNOVER INTENTION AND DEMOGRAPHIC FACTORS IN HOTEL BUSINESSES: A STUDY AT FIVE STAR HOTELS IN ISTANBUL

Begüm Dilara Emiroğlua , Orhan Akovab , Haluk Tanrıverdic

İstanbul University, İstanbul, 34126, Turkey

ABSTRACT

The study aims to determine the relation between the demographic factors (eg. age, gender, marital status, education) and the factors such as tenure, wage, position, working department and the turnover intention of employees working in hospitality industry. Surveys were used as the main data collection tool in the study. The study has been conducted in the various departments of five-star hotels in Istanbul and with a total of 297 hotel managers and staff from different ranks. In the data analysis of the study, t-test and ANOVA tests have been performed in order to determine the relationship between the demographic factors such as age, gender, marital status, education and the factors such as tenure, wage, position, working department and turnover intention. The result of the study indicates that the demographic factors such as age, gender, marital status, education, as well as the factors such as tenure, wage, position, working department are determinants for turnover intention.

KEYWORDS

Turnover intention, Hotel businesses, Demographic characteristics

1. INTRODUCTION

The staff working in tourism industry have to deal with the needs of customers as well as to cope with uncertain conditions at work (Kim et al., 2009: 612). That hotel industry is an industry based on human labor and that employees need to communicate with customers constantly which create interpersonal conflicts, staff stress and burnout (Hu and Cheng, 2010: 1337). High job stress that brings about a decrease in job commitment is considered as a factor that impacts the employee turnover considerably (Tiyce et al., 2013:127).

The major problem in the hotel industry is high turnover rate (Faldetta et al., 2013:584). Employee turnover rate in the hospitality industry often averages around 200 % to 300 % per year (Demir et al., 2007: 477). According to National Restaurant Association, the overall turnover rate in the restaurants and hospitality industry in U.S was 66.3 percent in 2014, a % 10 increase compared to 2010 (National Restaurant Association, 2015). Likewise, average annual employee turnover rate was % 64 percent in 2010 (Australian Tourism Labour Force Report, 2011). In a study conducted in Istanbul Hotels in Turkey in 2013, the average employee turnover rate at the hotel was found to be 21.9 percent (Hatipoğlu et al., 2013:31). Long working hours, low training, low salaries, stress, inadequate working conditions, poor management and lack of job opportunities, which is widely seen in the hotel industry, have led to leave employee jobs (Albattat and Som, 2013:64).

When an a employees leave their jobs, organization not only lose human capital but also endure the costs associated with high turnover of employees (Yang, 2008: 433). High employee turnover rate increase costs and affect productivity negatively (Lam et al., 2002:218). Therefore, the employee turnover is regarded as a negative factor that impacts quality of service, and mood of employee as well as profitability and other activities (Kim and Jogaratnam, 2010: 319). The effects of employee turnover on the organization are classified into two groups as direct and indirect effects. The direct effects may emerge when the costs of training and hiring new employees increase. Whereas, the indirect effects occur when the job satisfaction, organizational commitment, productivity decrease and subsequently the quality of service and reputation of the organization decrease (Faldetta et al., 2013: 584; Foreman 2009: 356; Hemdi and Nasurdin, 2004: 64). These adverse effects lead to degradation of quality of service subsequently and loss of customers (Hemdi and Nasurdin, 2004: 64).

One of the most important factors in the success of the tourism industry are employees. The competitiveness of tourism organizations and their survival depends on the quality of the services that they offer. In this case, tourism industry not only need to have trained staff that perform their roles effectively but should also need to retain their staff (Subramanian and Shin, 2013: 1751). For that reason, the factors that affect the turnover intention needs to be determined in order not to lose employees, to ensure their satisfaction and to meet their needs (Kim and Jogaratnam, 2010: 319).

Due to all these factors, turnover intention is of major importance in hospitality. Therefore, the study aims to determine the effect of the demographic factors and factors such as tenure, wage, position and working department on turnover intention.

2. LITERATURE REVIEW

2.1. Turnover Intention

Turnover intention is considered ″to be conscious and deliberate wilfulness to leave the organization″ (Tett and Meyer: 1993:262). Turnover intention is the intention to resign from his or her current job and the tendency to seek employment in other organization of an employee (Jung et al., 2012: 2148). In this regard, turnover intention represent that the personal estimate of the probability of an individual leaving work in the near future (Cho et al., 2009:374). Turnover intentions are divided into two categories: voluntary turnover and involuntary turnover (Albattat and Som, 2013: 63; Lam et al., 2003: 162). Voluntary turnover intention is the decision of an employee to leave the organization. Involuntary turnover intention, on the other hand, is the departure of an employee initiated by the employer (Lam et al., 2003: 162). Both the impact of the personnel's voluntarily turnover and personnel's involuntarily turnover to organization is enormous. Leave to job effects directly customers' perception of service quality degradation, the reduction of employees' job satisfaction and morale, increased of staff training costs and personal improvement (Cho et al, 2009:374)

The reasons of employee turnover can generally be divided into three categories as work-related factors (e.g. job satisfaction, wage, performance, organizational commitment), individual factors (e.g. age, education, gender, tenure) and external factors (e.g. unemployment rate, perceptions of employment, presence of trade union (Foreman, 2009: 357).

One of the most important factors of employee turnover is salary. Except the wage, the reasons of employee turnover are employees feeling the job or workplace is not what they expected, there is a mismatch between the job and person, there is too little coaching and feedback, there are few growth and career opportunities, employees feel devalued and unrecognized, employees feel stress from overwork and have a work/life imbalance, there is a loss of trust and confidence in senior leaders (Yazinski, 2009).

Despite being an undesired situation for organizations, turnover intention plays an important role since it is a determinant to minimize the negative impacts on organizational productivity (Kim et al., 2005: 175). High employee turnover rate is a major problem in hospitality industry (Lam et al., 2002: 218). Researchers have tried to uncover the reasons for employee turnover in the hospitality industry (Cho et al., 2009: 374).

Wood and Macaulay (1989) have conducted a study to determine why employees work in hospitality leave their jobs. According to them, reasons for employees' turnover are quality of supervision, lack of communication, working conditions, not to be compatible with company culture, low pay and few benefits, lack of clear definition of responsibilities, no direction on what to do (Gustafson, 2002:108). Getz (1994) highlighted that the hospitality industry involves common and undesirable characteristics such as low wages, poor working conditions, lack of occupational safety, lack of promotion opportunities and all these factors result in increasing staff turnover rate. Hai-Yan and Baum (2006) indicated that employees do not desire to work for a long time in hotels due to demanding responsibilities, low wage, and other uncertain promotion opportunities, which leads to low quality human resources.

Other researchers studying on the reasons of employee turnover in the hospitality industry found that the most important factors supporting the turnover intention are pay and benefits, job satisfaction as a whole, relationship with the coworkers, whether pre-employment expectations are met or not, tenure, job performance, internal motivation, job status (hourly vs. salary), job stress, complexity (Pizam and Thornburg, 2000), company factors (etc. job environment, company sub-culture, company decision-making, management style, owner's financial status), compensation and promotion channels (etc. higher paid positions, bonuses and incentives, career promotion systems), personal feelings, the nature of hotel industry, work content (etc. lack of motivation, excessive information processing, the monotony of daily work, change in philosophy of business operations (Yang et al., 2012), low salary, poor working conditions, job stress, injustice (Albattat and Som, 2013).

Some of the studies on turnover intention in tourism industry found out that demographic variables have an effect on turnover intention. Age was found to be a determinant on turnover intention, and a negative relationship between age and turnover intention was found; it was also argued that the turnover intention of older employees is generally low (Carbery et al., 2003; Ghiselli et al., 2001; Kim et al., 2010; Karatepe et al., 2006; Pizam and Thornburg, 2006;). Additionally, the studies on gender and turnover intention indicated that the factor of gender is a determinant for turnover intention (Carbery et al., 2003; Karatepe et al., 2006; Uludağ et al., 2011). In some of the studies on the relationship between educational level and turnover intention, a positive relationship was found between these factors and it was concluded that the people with low educational level have a lower turnover intention (Carbery et al., 2003), whereas some of the studies found a negative relationship (Karatepe et al., 2006). Marital status was also revealed to be a determinant for turnover intention (Carbery et al., 2003).

Various studies on the turnover intention aimed to reveal the effect of the factors such as wage, tenure, position, working department other than the demographic characteristics on turnover intention and the relationship between them. Wage was found to be associated with turnover intention and it was concluded that low wage strengthens turnover intention (Albattat and Som, 2013; Carbery et al., 2003; Ghiselli et al., 2001; Pizam and Thornburg, 2000). Tenure was also found as one of the major determinant factors of turnover intention and it was revealed that the people with high tenure have generally have lower turnover intention (Karatepe et al.,, 2006; Nadiri and Tanova, 2010; Uludağ et al., 2011). The factor of position was also revealed to be a determinant for turnover intention and generally it was revealed that people with low position have generally higher turnover intention more than people with high position (Carbery et al., 2003; Ghiselli et al., 2001; Kim et al., 2010). Likewise the position, the factor of working department were also revealed to be a determinant for turnover intent and it was concluded that the factor of working department affect to turnover intention (Gallardo et al., 2010; Pizam and Thornburg, 2000). Moreover, in many studies, job satisfaction was found to be one of the determinants for turnover intention and it was indicated that those with low job satisfaction have a higher turnover intention (Carbery et al., 2003; Hemdi and Nasurdin, 2004; Kim et al., 2005; Zopiatis et al., 2014). Organizational commitment and organization support were also found to be one of the determinants of turnover intention and there was a negative relationship between these aspects and turnover intention (Carbery et al.,, 2003; Cho et al., 2009; Yang, 2008).

3. METHODOLOGY

3.1. Purpose of Study

This study is deemed important in order to determine the demographic factors and factors such as tenure, wage, position and working department that have an impact on the turnover intention of the employees working at five-star hotels in Istanbul and to identify the attitudes of staff towards their turnover intention. That said, the research hypotheses were determined as follows:

H1: There is a significant difference between the demographic factors and turnover intention.

H2: There is a significant difference between the tenure and turnover intention.

H3: There is a significant difference between the position and turnover intention.

H4: There is a significant difference between the wage and turnover intention.

H5: There is a significant difference between the working department and turnover intention.

3.2 Importance of the Study

One of the major problems in the contemporary tourism industry is the high turnover rate of staff. There are direct and indirect factors of employee turnover such as increasing costs of training and hiring new employees, decreasing employee productivity, satisfaction, commitment and mood, and subsequently degradation of quality of service and poor organizational image. Therefore, the factors that impact employee turnover should be determined to reduce the turnover of employees, which will affect employee productivity, satisfaction and commitment and also have a contribution in organization productivity, service quality and organization image in the long term. The study considering all these factors has an important place in determining the demographic factors and factors such as tenure, wage, position and working department that impact the turnover intention of employees working in five-star hotels in Istanbul.

3.3. Population and Sampling of the Study

The population of the study is composed of employees working in five-stars hotels in Istanbul. The sample of the study consists of 297 managers and staff working in a total of ten five-star hotels that were selected conveniently among hotels in Istanbul. Due to time and cost constraints, a total of eight five-star hotels were selected and 400 survey forms were distributed by visiting these hotels. A total of 310 forms were received and 13 of them were not taken into consideration since they were incomplete. 297 surveys were used during data analysis.

3.4. Method of the Study

The study was conducted between September and February 2015. A scale of three questions and with a single factor developed by Mobley, Griffeth, Hand ve Meglino (1979) was used to evaluate the turnover intention of the participants of the survey. A five-point Likert scale was utilized in the survey form. The evaluation options in regard to the concepts of turnover intention are listed as follows: (1) I definitely disagree, (2) I disagree, (3) I neither agree nor disagree, (4) I agree, (5) I definitely agree. The internal reliability of the scale of turnover intention was found as 0.980. The findings were analyzed by means of t-tests and Anova tests in order to determine the demographic factors and factors such as tenure, wage, position, working department that might have an impact on the turnover intention of the staff who participated in the study.

4. FINDINGS

Table 1 indicates the distribution of the participants in terms of demographic characteristics. 48.8% of the participants of the study are male, and 51.2% are female. 75.8% of the participants are married whereas 24.2% are single. 44,1% of them are between 18-35 years old, 49% are between 36-55 years old, 6.7% are 56 years old and older. In terms of education level, a high majority of the participants as much as 54.9% are university graduates. In terms of tenure, a high majority of them as 38.0% have been working in their current job for 5-7 years. Furthermore, in terms of their working department, the majority of the participants are staff (30.6%) work in the food and beverage department, 19.2% work in the housekeeping department, and 15.2% work in the kitchen and front office departments. In terms of position, the majority (79.8%) are staff without managerial responsibilities while in terms of wage, 80.5% of them have a wage between 1500-1999 Turkish Liras.

Table 2 indicates the distribution in regard to frequency and percentage of the answers of the participants. In terms of the answers of the participants on turnover intention, 62.0% of them stated that they do not think of leaving from the organization that they are currently working in; 59.5% stated that they do not actively seek for alternative employment, and 62.3% stated that they would not resign from the organization at the earliest possible opportunity. The turnover intention of the participants was found to be at a moderate level (2,6397).

4.1 Testing the Hypotheses

H1 hypothesis was tested according to the variables such as gender, age, education level and marital status in order to determine the effects of the demographic factors on turnover intention.

Table 1. Descriptive Information on the Demographic Information of the Participants

GENDER	(f)	(%)	MARITAL STATUS	(f)	(%)	EDUCATION	(f)	(%)	POSITION	(f)	
Female	145	48.8				Primary School	30	10.1	Manager	28	(%)
Male	152	51.2	Married	225	75.8	Secondary School	23	7.7	Assistant Manager	14	9.4
TOTAL	297	100	Single	72	24.2	High School	64	21.5	Chief	18	4.7
			TOTAL	297	100	University	163	54.9	Staff	237	6.1
						Other	17	5.7	TOTAL	297	79.8
						TOTAL	297	100			100

AGE	(f)	(%)	TENURE	(f)	(%)	WAGE	(f)	(%)	DEPARTMENT	(f)	(%)
18-25	57	19.2	1 year and less	20	6.7	1500-1999 TL	239	80.5	Front Office	45	15.2
26-35	74	24.9	2-4 years	72	24.2	2000-2499 TL	27	9.1	Housekeeping	57	19.2
36-45	118	39.7	5-7 years	113	38.0	2500-2999 TL	21	7.1	Food and	91	30.6
46-55	28	9.4	8-10 years	76	25.6	3000-3499 TL	7	2.4	Beverage	45	15.2
56 and older	20	6.7	10 years and more	16	5.4	3500 and above TL	3	1.0	Kitchen	12	4.0
TOTAL	297	100	TOTAL	297	100	TOTAL	297	100	Accounting	8	2.7
									HR	37	12.5
									Technical Service	2	0.7
									Purchasing	297	100
									TOTAL		

Table 2. Descriptive Information on the Responses and the Turnover Intention Levels of the Participants

	I definitely disagree/ I disagree		I neither agree nor disagree		I definitely agree/ I agree		N	$\bar{X}$	S
	f	%	f	%	F	%			
I am thinking of resigning from this organization.	184	62.0	3	1.0	110	37.0	297	2.6431	1.57060
I am actively looking for alternative employment.	178	59.5	2	0.7	117	39.4	297	2.6599	1.54947
I will resign from this organization at the earliest possible opportunity.	185	62.3	2	0.7	110	37.0	297	2.6162	1.56853
Turnover Intention							297	2.6397	1.53219

Table 3. Independent Samples T-Test Indicating the Difference of Turnover Intention among the Employees in terms of Gender

Gender	N	X̄	S	df	t	P
Male	145	2,3080	1.46043	294,996	-3,727	,000*
Female	152	2,9561	1.53674			

*p<0,05

A significant difference between gender and turnover intention (p<50,5) A significant difference between gender and turnover intention (X̄ =2,9561). This result is consistent with previous studies that found females have a higher turnover intention than males (Carbery et al.,, 2003; Cotton and Tuttle, 1986; Lambert, 2006).

Table 4. Independent Samples T-Test Indicating the Difference of Turnover Intention of the Employees in terms of Marital Status

Marital Status	N	X̄	S	df	t	P
Married	225	2,2696	1,42727	134,839	-8,691	,000*
Single	72	3,7963	1,25292			

*p<0,05

It was found a significant difference between marital status and turnover intention (p<0,05) and marital status was found based on the t-test analysis. In this study, it was concluded that single people have a higher turnover intention than married people (X̄=3,7963)This result is consistent previous studies made by Carbery et al., 2003 and Cotton and Tuttle, 1986

As can be seen in table 5, it was found a significant difference between age and turnover intention (p<0,05) and factor of age was found to be a determinant on turnover intention. In research, older people who is located in the age groups 46-55 and 56 years and over have lower turnover intention than young people who is located in other age groups (X̄= 3,9123, 3,3955, 1,4414, 1,2024, 1,0000). Previous studies have found similar results and they support this finding that older people have lower turnover intention than young people (Chen et al., 2010; Cotton and Tuttle, 1986; Iverson and Deery, 1997; Lambert, 2001; Lambert, 2006; Martin and Roodt, 2008). As a result of the analyses, between in 18- 25 and 26-35 age groups, between in 36-45 and 46-55 age groups and finally between in 46-55 and 56 years and over age groups, it wasn't have any difference, it was observed difference among in other age groups except for this age groups

Table 5. ANOVA Table Showing the Difference of Turnover Intention of the Employees in terms of Age

Age	N	$\bar{X}$	S	Difference	df	F	P
18-25 (1)	57	3,9123	1,15909	1-3 1-4 1-5			
26-35 (2)	118	3,3955	1,11854	2-3 2-4 2-5			
36-45 (3)	74	1,4414	,08414	3-1 3-2 3-5	4	65,146	,000*
46-55 (4)	28	1,2024	,07540	4-1 4-2			
56 age and + (5)	20	1,0000	,00000	5-1 5-2 5-3			

*p<0,05

Table 6. ANOVA Table Showing the Difference of Turnover Intention of the Employees in terms of Education Level

Education Level	N	$\bar{X}$	S	Difference	df	F	P
Primary School (1)	30	1,5333	1,01935	1-4 1-5			
Secondary School (2)	23	1,6087	,80184	2-4 2-5			
High School (3)	64	2,0365	1,06924	3-4 3-5			
University (4)	163	3,0593	1,56826	4-1 4-2 4-3 4-5	4	21,209	,000*
Other (5)	17	4,2353	1,21772	5-1 5-2 5-3 5-4			

*p<0,05

As a result of the analyses, a significant difference between education level and turnover intention was also found(p<0.05) People who have lower levels of education such as primary school, secondary school, high school have lower turnover intention compared to people who have higher levels of education such as university and higher ($\bar{X}$= =1,5333,

1,6087, 2,0365, 3,0593, 4,2353). This finding is similar to that of the earlier research which found that people who have low levels of education have low turnover intention (Chen et al., 2010; Iqbal, 2010; Khatri, 2001; Lambert, 2006). There were no significant differences found among primary, secondary and high school graduates.

According to the results, there was a significant difference between the demographic factors and turnover intention, and hypothesis H1 was not rejected.

Table 7.ANOVA Table Showing the Difference of Turnover Intention of the Employees in terms of Tenure

Tenure	N	$\bar{X}$	S	Difference	df	F	P
1 year and less (1)	20	4,3500	,81273	1-3 1-4 1-5			
2-4 years (2)	72	3,8194	1,19589	2-3 2-4 2-5			
5-7 years (3)	113	2,5870	1,50018	3-1 3-2 3-4 3-5	4	50,907	,000*
8-10 years (4)	79	1,4167	,60154	4-1 4-2 4-3			
10 year and more (5)	13	1,3750	,46944	5-1 5-2 5-3			

*p<0,05

As can be seen in table 7, there is a significant difference between tenure and turnover intention (p<0.,05) and hypothesis H2 was not rejected. In this research, it was concluded that people who have tenure of 5-7 years, 8-10 years and 10 years and over have lower turnover intention compared to people who have relatively short tenure such as one year and less, 2-4 years. ($\bar{X}$=4,3500, 3,8194, 2,5870, 1,4167, 1,3750). This finding corresponds with previous studies that found people who have long tenure have low turnover intention more than people who have short tenure (Cotton and Tuttle, 1986; Lambert, 2006; Martin and Roodt, 2008). No significant difference was found between in 1 year and less and 2-4 years tenure groups as well as among 8-10 years and 10 years and more tenure groups

Table 8. ANOVA Table Showing the Difference and Turnover Intention of the Employees in terms of Position

Position	N	$\bar{X}$	S	Difference	df	F	P
Manager (1)	28	1,1786	,35697	1-3 1-4	3	17,486	,000*
Assistant Manager (2)	14	1,2619	,43713	2-3 2-4			
Chief (2)	18	2,3704	1.17651	3-1 3-2			
Staff (3)	237	2,9142	1.54031	4-1 4-2			

*p<0,05

A significant difference between position and turnover intention was also found (p<0,05) and hypothesis H3 was not rejected. Employees who have higher position such as managers and assistant managers have lower turnover intention than people who have low position such as chief and baseline staff. ($\bar{X}$=,1786, 1,2619, 2,3704, 2,9142). This finding is consistent with previous studies that found people who have higher position have lower turnover intention compared to people who have lower positions. (Abdullah et al., 2010; Radzi et al., 2009). Moreover, there were no significant differences between in manager-assistant manager position groups and between in chief-baseline staff position groups

Table 9. ANOVA Table Showing the Difference of Turnover Intention of the Employees in terms of Wage

Wage	N	$\bar{X}$	S	Difference	df	F	P
1500-1999 TL (1)	239	2,9010	1,54090	1-2 1-3 1-4 1-5	4	11,553	,000*
2000-2499 TL (2)	27	2,0370	1,09908	2-1 2-3 2-4 2-5			
2500- 2999 TL (3)	21	1,1746	,35931	3-1 3-2			
3000-2499 TL (4)	7	1,1429	,37796	4-1 4-2			
3500 and more TL (5)	3	1,000	,00000	5-1 5-2			

*p<0,05

A significant difference wage and turnover intention was found (p<0,05) and hypothesis H4 was not rejected. Employees who are paid between 1500-1999 TL and 2000-2499 TL have relatively lower wage groups have higher turnover intention than people who have higher wage; 2500 TL and over. This finding corroborates with previous studies that argue staff who have low wage have a higher turnover intention. (Albattat and Som, 2013; Carbery et al, 2003; Pizam and Thornburg, 2000)

A significant difference between working department and turnover intention was also found ($p<0,05$) and hypothesis H5 was not rejected. The staff working front office, food and beverage and kitchen had a higher turnover intention than those working in other departments. It was concluded that particularly the staff working in the front line (especially in front desk and food and beverage) had a higher turnover intention than those working in the back of the office (human resources, accounting, purchasing). Furthermore, it was found that the kitchen staff in back office are quite dissatisfied with their job. In the study also found that staff working in the housekeeping and technical service had a lower turnover intention

4. DISCUSSION AND CONCLUSION

This study aims to determine the effect of the demographic factors and factors such as tenure, wage, position and working department on turnover intention, the turnover intention of the employees was found to be at moderate level.

It was concluded that the turnover intention varies based on demographic factors and such as gender, marital status, age, education level and factors such as tenure, wage, position and working department. As concluded in earlier studies (Carbery et al., 2003; Karatepe et al., 2006; Uludağ et al., 2011), gender has been found to be a determining factor turnover intention. The study demonstrated that there the female employees have a higher turnover intention than male employees. It can be concluded that this situation is a result of the role and responsibilities assumed for the female employees within their families and the limited career opportunities. Griffeth et al. (2000:484) indicated that this situation is a consequence of the primary responsibilities of women as traditional household chores and child care, and that childbearing requires women to leave their paid employment. While other researchers who study this issue report that this result stems from the responsibility of women in their families (Keith and McWilliams, 1995; Sicherman, 1996), the availability of limited opportunities to women for improvement (Stroh et al., 1996) and the poor commitment to their jobs (Chaudhury and Ng, 1992).

Table 10. ANOVA Test Showing the Difference of Turnover Intention of the Employees in terms of Working Departments

Working Department	N	X̄	S	Difference	Working Department	N	X̄	S	Difference	df	F	P
Front Office (1)	45	3,6296	1,41282	1-2 1-5 1-6 1-7	**Purchasing (5)**	2	1,0000	,00000	8-1 8-2 8-3 8-4	7	27,647	,000*
Housekeeping (2)	57	1,5380	,79870	2-1 2-3 2-4	**Technical Service (6)**	37	1,4234	,60680	7-1 7-3 7-4			
Food and Beverage (3)	91	3,1868	1,44804	3-2 3-5 3-6 3-7 3-8	**Kitchen (7)**	45	3,6074	1,35641	4-2 4-5 4-6 4-7 4-8			
Accounting (4)	12	1,3333	,86457	5-1 5-3 5-4	**Human Resources (8)**	8	1,2500	,46291	6-1 6-3 6-4			

*p<0,05

The turnover intention was also found to differ based on marital status and the single employees were observed to have a higher turnover intention compared to married employees. The reason of this may be due to the financial responsibilities assumed by the married employees in their families. This finding support the findings of Carbery et al. (2010). Carbery et al., (2010) reached the conclusion that the turnover intention of married employees was lower due to their financial obligations.

The turnover intention was also found to differ based on age and there was not any difference in the turnover intention of the people with similar age groups. Similar to the earlier research in general (Carbery et al., 2003; Ghiselli et al., 2001; Karatepe et al., 2006; Kim et al., 2010; Pizam and Thornburg, 2006), it was found that the older people had a lower turnover intention. The reason that the older people had a lower turnover intention can attributed to the fact that they do not have similar thoughts and expectations compared to younger people. The reasons for this situation have been discussed differently by various researchers. Martin and Roodt (2008) supported that old people tend to have an intention to stay in the organization for a longer time since they are not equal to a more qualified person and invested more in the organization. Cotton and Tuttle (1986) stated that as people get older, they become familiar with their jobs and since they have a close relation with their friends, the turnover intention decreases. Lee (2013) regarded this situation as the loss of work productivity of old people. Spector (1997) reported that since old people use their skills better, work under better conditions, utilize advancement and promotion opportunities and appreciate fringe benefits, they have low expectations compared to young people. Furthermore, old people are not classified in the same generation with young people. A high majority of the participants in this study are from the generation Y. This may have an effect on their turnover intention. Solnet and Hood (2008) stated that generation Y constitutes the majority of the staff working in the tourism industry, that these staff mostly work at front office and that they are considerably different from the previous generations in terms of work-related characteristics, values and attitudes. A study on hotel employees by Lub et al., (2012) reported that generation Y has a high expectations in terms of work-life balance, autonomy and occupational safety and a higher turnover rate, that the expectations of generation X and Y are different from each other and that generation Y value different aspects of the work. It was revealed that the expectations of generation Y employees in hospitality are frequently incompatible with their opinions and the reality of the industry (Barron et al., 2007; Cairncross and Buultjens, 2007).

Turnover intention was also found to differ based on various education levels. Similar to the earlier research (Carbery et al., 2003; Karatepe et al., 2006), it was concluded that those with low education level has a lower turnover intention. Furthermore, the turnover intention of those with high education level was higher in the study. This may be resulted from that the people with high education level have more expectations compared to those with low education level. Other researchers who study on this subject have reached similar conclusions. Chen et al., (2010) reported that the people with high education level have higher expectations in terms of financial rights, benefits and audit compared to those with low education level. Iqbal (2010) stated that those with high education level have higher expectations towards their current employers, which means that the fulfilment of their needs is much more difficult. Lambert (2006) supported that more educated people desire to obtain more job opportunities compared to uneducated people and have a wider view/understanding that creates new challanges, which results in increasing expectations and that a person, in such a situation, might regard the organizations as a prison.

Turnover intention also differ based on various wage groups. This conclusion is parallel to that of earlier research (Albattat and Som, 2013; Carbery et al., 2003; Ghiselli et al., 2001; Pizam and Thornburg, 2000) and it was concluded that low wage affects employee churn.

The turnover intention was also found to differ based on tenure similar to the conclusions of earlier research (Karatepe et al.,, 2006; Nadiri and Tanova, 2010; Uludağ et al., 2011) and the study concluded that those with long tenure had a lower turnover intention. While there was no difference between the groups with similar tenure, a difference between other groups was observed. Those with higher tenure had a lower turnover intention may be a result of a staff who does not think of resigning from the employment due to their high investment in the organization and the intensity of his or her interaction with other people as he or she spends more time in that organization. The result of the study supports Becker's (1960) theory. Becker (1960) advocated that an employee invests in the organization over time (e.g. retirement, pensions, pay raises, benefits, stocks, position) and these investments bond the individual to the organization. Becker calls these investments as sunk costs. As reported by Becker, as age and tenure increases, the individual bonds to the organization and the possibility of resigning decreases. Other researchers supported that as the individual spends more in the organization, he or she builds strong interpersonal ties by means of repeated interactions with other people and that these strong interpersonal ties lead to a decrease in turnover intention (Blau and Khan 1981; Griffeth et al., 2000).

The turnover intention was also found to differ based on position. Similar to previous studies (Abdullah et al., 2010; Radzi et al., 2009), it was found that those with high position level had a lower turnover intention. It was observed that particularly those working in the position of manager and assistant manager had a lower turnover intention compared to those in the position of chief and staff. This may a result of monetary incentives and decision-making responsibilities of managers. Mcguire et al., (2003) and Bonn and Forbringer (1992) revealed that wage and incentives are effective for employees to stay in the organization. Salami (2008) advocated that since those working in higher positions have more responsibilities in the organization, they show low turnover intention. Similarly, Khatri et al., (2001) reported that when managers provide fair and consistent organizational policies and practices in regard to the procedure to make a value available in the organization such as promotion, termination, performance evaluation, bonus or assume similar tasks, they tend to have a lower turnover intention. This is also supported by the hospitality researchers. For instance, Radzi et al.,(2009) stated that the managers are mainly responsible for the tasks such as wage satisfaction, work schedule, workload, rewards, and job responsibilities; therefore, they do not think of other alternatives outside the organization due to these responsibilities and they tend to have lower turnover intention. That those with low position leave the organization may be resulted from low wages and limited promotion opportunities. Hai-Yan and Baum (2006) found that the employees do not desire to work for a long time in the hotel due to labor demand, low wages and uncertain promotion opportunities.

The turnover intention was also found to differ based on working department. There are scant research on department and turnover intention in the literature. A study on the staff working in hospitality by Pizam and Thornburg (2000) concluded that while the staff with highest absenteeism rate work in the housekeeping, and respectively in the food and beverage, sales, front office, technical service, safety and human resources departments. In our study, the turnover intention of the staff working particularly in the front office, food and beverage departments was higher compared to that of those in the back office such as human resources, accounting, purchasing, etc. departments. It can be stated that this is a result of more workload of the staff working in the front office and more interaction with the customers. Moreover, the study found that the staff working in the housekeeping and technical service departments had a lower turnover intention, which may stem from different expectations of people depending on their education level. However, contrary to the finding by Pizam and Thornburg (2000), the turnover intention of those in the housekeeping department was found low in this study

That said, it was concluded that demographic factors and the factors such as tenure, wage, position and working department have an influence over the turnover intention of employees. Therefore, the hospitality industry need to understand these demographic factors and the factors such as tenure, wage, position and working department to take measures to decrease turnover intention. Based on the findings of the study, human resource practices can be applied successfully to influence employees' beliefs about the commitment to the organization. Recent empirical research on the general business literature revealed that human resource practices can contribute to organizational commitment of employees. Human resource strategies should be applied to increase the perceived organizational support of employees, to reduce their intention to leave their job and to retain them. Such efforts help increase the perceived organizational support from workers and strengthens employees' knowledge and skills (Cho et al., 2009: 380). The higher the perceived organizational support of individuals, the less intention to seek alternative jobs and probability of accepting these jobs. Human resources practices that recognize the contributions of employees and investing in them (eg. career opportunities, awards, participation in decision making) contributes to an improvement in employees perceptions of organizational support (Allen et al., 2003: 102).

The results of this study have important implications that it explores the turnover intention of the staff in hospitality based on demographic factors and the factors such as tenure, wage, position and working department. In addition, the study is restricted to the staff in five-star hotels in Istanbul. Provided that the study was performed in a different scale or different areas of tourism industry or in the tourism businesses with different customer profiles or product concept, it could reach different conclusions. Hence future studies might look into those areas.

REFERENCES

1. Abdullah, R.B., Alias, M.A.M., Zahari, H., Karim, N.A., Abdullah, S.N., Salleh, H., Musa, M.F. (2010). The Study of Factors Contributing to Chef Turnover in Hotels in Klang Valley, Malaysia. Asian Social Science, 6 (1): 80-85

2. Albattat, A. R.S., Som, A.P.M. (2013). Employee Dissatisfaction and Turnover Crises in the Malaysian Hospitality Industry. International Journal of Business and Management, 8 (5): 62-71

3. Allen, D.,Shore, L.M., Griffeth, R.W. (2003). The Role of Perceived Organizational Support and Supportive Human Resource Practices in The Turnover Process. Journal of Management, 29 (1) 99–118.

4. Australian Tourism Labour Force Report (2011). Australian Tourısm Labour Force Report Labour Force Profile (Part 1), Deloitte Access Economics.

5. Barron, P., Maxwell, G., Broadbridge, A., Ogden, S. (2007), Careers in Hospitality Management: Generation Y's Experiences and Perceptions. Journal of Hospitality and Tourism Management, 14 (2): 119-128.

6. Becker, H. (1960). Notes on The Concept of Commitment. American Journal of Sociology, 66, 32-42

7. Blau, F. D. and Khan, L.M. (1981). Race and Sex Differences in Quits by Young Workers. Industrial and Labor Relation Review 34 (4): 563-577.

8. Bonn, M. A., Forbringer, L. (1992) Reducing Turnover in The Hospitality Industry: An Overview of Recruitment, Selection and Retention. International Journal of Contemporary Hospitality Management, 11(1), 47-63.

9. Cairncross, G., Buultjens, J. (2007), Generation Y and Work in The Tourism and Hospitality Industry: Problem? What problem?

Occasional paper No. 9, Centre for Enterprise Development and Research, Southern Cross University, Tweed Heads, Australia.

10. Carbery, R, Garavan, T.N., O'Brien, F., McDonnell, J. (2003). Predicting Hotel Managers' Turnover Cognitions. Journal of Managerial Psychology, 18 (7): 649-679
11. Chaudhury, M., Ng, I. (1992). Absenteeism Predictors: Least Squares, Rank Regression, and Model Selection Results. Canadian Journal of Economic, 25: 615-635.
12. Chen, Y.C., Kuo, C., Cheng, W.W.,Hsai, H.,Chien, C.Y. (2010). Structural Investigation of the Relationship Between Working Satisfaction and Employee Turnover. The Journal of Human Resource and Adult Learning, 6 (1): 41-50
13. Cho, S., Johanson, M.M., Guchait, P. (2009). Employees Intent to Leave: A Comparison of Determinants of Intent to Leave Versus Intent to Stay. International Journal of Hospitality Management, 28: 374-381
14. Cotton, J.L., Tuttle, J.M. (1986). Employee Turnover: A Meta-Analysis and Review with Implications for Research. Academy of Management Review 11(1): 55-70.
15. Demir, C., Çolakoğlu, Ü.,Güzel, B. (2007). Relationship Between Employee Turnover and The Location of Hotels: The Case of Kuşadası and Izmir in Turkey. Journal of Yasar University, 2 (5): 477-487.
16. Faldetta, G., Fasone, V., Provenzano, C. (2013). Turnover in the Hospitality Industry: Can Reciprocity Solve the Problem. Revista de Turismo y Patrimonio Cultural, 11(4):583-595.
17. Foreman, W.T. (2009). Job Satisfaction and Turnover in The Chinese Retail Industry. Chinese Management Studies, 3(4): 356-378
18. Gallardo, E., Sánchez-Cañizares, S.M., Lôpez-Guzmán, T., Jesus, M.M.N. (2010). Employee Satisfaction in The Iberian Hotel Industry The Case of Andalusia (Spain) and The Algarve (Portugal).

International Journal of Contemporary Hospitality Management, 22 (3): 321-334.

19. Getz, D. (1994). Residents' Attitudes Toward Tourism: A Longitudinal Study in Spey Valley, Scotland. Tourism Management, 15(4), 247-258

20. Ghiselli, R.F., Lopa, J.M.L., Bai, B. (2001). Job Satisfaction, Life Satisfaction, and Turnover Intent. Cornell Hotel and Restaurant Admınıstratıon Quarterly, 42 (2): 28-37.

21. Griffeth, R.W., Hom, P.W., Gaertner, S. (2000). A Meta-Analysis of Antecedents and Correlates of Employee Turnover: Update, Moderator Tests, and Research Implication for the Next Millennium. Journal of Management 26 (3): 463-488.

22. Gustafson, C.M. (2002). Employee Turnover: A Study of Private Clubs in The U.S.A. International Journal of Contemporary Hospitality Management, 14 (3): 106-113.

23. Hai-Yan, K., Baum, T. (2006) Skills and Work in The Hospitality Industry Sector: The Case of Hotel Front Office Employees in China. International Journal of Contemporary Hospitality Management, 18(6): 509-518.

24. Hatipoğlu, B., Alvarez, M.D., İnelmen, K., Ünalan, D. (2013). İstanbul Otellerinin 360° Değerlendirmesi Araştırma Proje Raporu 2013. TUROB, 1- 95.

25. Hemdi, M.A., Nasurdin, A. M. (2004). A Conceptual Model of Hotel Managers' Turnover Intentions: The Moderating Effect of Job-Hopping Attitudes and Turnover Culture. Tourism Educators Association of Malaysia, 1(1), 63-76

26. Hu, H.H.S., Cheng, C.W. (2010). Job Stress, Coping Strategies, and Burnout Among Hotel Industry Supervisors in Taiwan. The International Journal of Human Resource Management, 21 (8): 1337-1350.

27. Iqbal, A. (2010). An Empirical Assessment of Demographic Factors, Organizational Ranks and Organizational Commitment. International Journal of Business and Management, 5 (3): 16-27

28. Iverson, R.D., Deery, M. (1997). Turnover Culture in The Hospitality Industry. Human Resource Management Journal. 7 (4):71-82

13

THE DETERMINANTS OF ONLINE HOTEL RESERVATIONS AMONG UNIVERSITY STAFFS

Abdul Rahim Abu Bakar

Universiti Utara Malaysia, Malaysia, arab1024@uum.edu.my Fariza Hashim, PhD, College of Business, Universiti Utara Malaysia, Malaysia,

ABSTRACT

To compete in the current business environment, hoteliers need to cut down their distribution costs and adopt the electronic distribution channels such as the Internet. However, Malaysian hotels have been relatively slow in adapting this due to the current preference of using traditional method of room reservation by hotel patrons. Based from the literature review of the Internet users, a public university was chosen to represent one of the profiles of Internet users. A sample of 193 respondents was collected and the findings indicated that profiles conformed to previous studies. In adopting online reservation, online "bookers" would look for fast, reliable and convenient system. Price is a key motivator to purchase online. The findings are important in encouraging more hoteliers to adopt and implement online hotel reservations as part of their main sales mechanism.

1. INTRODUCTION

The travel sector is rated among the top three product or service categories purchased via the Internet (Eric et.al, 2006)[1]. According to research carried out by Jupiter Research, the number of people who have bought travel product over the Internet is likely to double from 18.6 million in 2001 to 38.6 million by 2007. One sector within the travel industry that has been transformed by the ICT developments is the hospitality section. It is said that to compete effectively in the current business environment, hoteliers need to adopt the electronic distribution channels. In addition, research results indicate that a growing trend in rooms booked via the Internet is not emerging from existing customer bases, thereby implying that the Internet is able to attract new market segments (Connolly et al.,1998). [2]

In response to this e-business opportunity, most hotels have already established their own websites to facilitate promotion and reservations on the web Starkov and Price (2007)[3] predicted that at least one-third of all hotel bookings will be completed online in 2010. With respect to Malaysia, in 2005 45% of online purchases in Malaysia were related to travel related items such as flight tickets and hotel bookings. With the Internet users currently at 14.9 million which is 60 per cent of the country's population; an increase of 302.8 per cent increase compared to the year 2000 (Internet World Stats,2008)[4], this represents an attractive market potential. Nevertheless, while the proportion of online reservations is increasing, only 64% of hospitality firms currently handle such transactions (Client and Warner,2001)[5]. According to Cobanoglu (2001)[6], travelers still use travel agents as their favorite reservation resource followed by toll-free reservation numbers, and then calling the hotel directly Peterson et. al (1997).[7] find that consumers are able to gather information about products and services on the Internet; however, the actual distribution of goods and services is likely to be affected by the Internet

Although ICT experts predict that within several years the Internet will be one of the most important sources for hotel reservations and services (Cline and Warner, 2001)[5], it is intriguing that the current customers still prefer using traditional methods which are more costly and inefficient. In the Malaysian hotel sector for instance, the Malaysia Association of Hotel (MAH) is facing problems of insufficient fund to support the online reservation system due to lack of respond from many other hotels. This is despite the system is said to cuts costs, attracts affluent customers and lessens the dependency on more traditional and expensive channels.

The lukewarm response from hoteliers could be due to the current sluggish online bookings from travelers. Although travelers are increasingly sophisticated in their technological interactions, they may avoid the online reservation system if they feel uncomfortable, even when the benefits are obvious (Meuter et al., 2003)[8]. Hence, despite the growing online market, hotels are still searching for an efficient way to persuade travelers to reserve rooms directly through the hotel online reservations. To add salt to injury, progress in related research has been somewhat lagging (Pernsteiner and Rauseo, 2000)[9]. Earlier review of existing literature indicated that too little is known of the drivers of online consumer behavioural intentions (e.g. Grewal et al., 2004)[10]. While earlier researchers have focused on online service quality (e.g. Zeithaml et al., 2002)[11], a number of commentators have observed that online behavioural intentions while central, have somewhat been neglected (Goode and Harris, 2007)[12]. The gap in knowledge is further compounded by the dearth of research into factors that may moderate the linkage between such antecedents and behavioral intentions (e.g. Jones et al., 2000)[13].

Although there have been efforts to study hotel customers' online purchase intention, much of the research has focused on Western customers and cultures. Empirical studies pertaining to ecommerce across the Asia-Pacific Rim remain rare (Wu, Zhu and Xu, 2000)[14]. In response to these gaps, this paper identifies and analyses the important factors that affect hotel customers' online reservation intention in relation with their profiles and Internet behaviour usage patterns. The understanding of the online purchase intention will help to increase the share of online hotel sales total room revenues. Consequently, this would encourage more hoteliers to adopt and implement online hotel reservations as part of their main sales mechanism. Online reservations can reduce marketing costs and increase sales by efficiently identifying potential online customers. Additionally, online travel retailers can increase their revenues by improving marketing strategies and provide better service to attract and retain more online customers if they know which factors significantly increases the probability of online booking. Only through this understanding will marketers be able to develop strategies and tactics to attract and maintain these customers.

2. LITERATURE REVIEW

2.1 Demographic characteristics and Internet Usage Behaviour

Education level, age, income, and occupation have been found to be significantly different among Internet users. Bonn et al. (1998)[15] study showed that those who use the Internet as a travel information-gathering tool are likely to be more educated, younger, with higher household incomes, use commercial lodging accommodations while traveling, tend to travel by air and spend more money on travel-related expenses. Weber and Roehl's (1999)[16] study shows similar results. They found that online travel purchasers are more likely to be people who are aged 26-55, with higher incomes, with higher status occupations, and have more years of experience with the Internet than those who do not search or purchase online. The Asian demographic patterns of online purchasers tend to mirror their Western profiles. According to studies carried out by Technowledge Asia in 1999 and 2000, cybershoppers in the four Asian regions (Singapore, Hong Kong, Taiwan and Malaysia) were found to be mainly males, in the age group of 26 to 35 years and better educated.

Hence, it has been shown that an individual's educational level may affect their susceptibility to attitude change. Intelligent people understand complex messages better, and as a result, persuasion by complex messages is more likely. In addition, an individual's educational level is also partially a measure of their socioeconomic status. Weber and Roehl (1999)[16] study found that Internet "bookers" were more likely to have used the Internet for 4 years. Other researchers confirmed that Internet bookers spend more time online per week than those booking offline (Weber and Roehl ,1999)[16].

Finally, the intention to shop online is also influenced by consumers' Internet shopping history (Shim et al., 2001)[17]. It is demonstrated by past research findings that prior online shopping experiences have a direct impact on Internet shopping intentions (Weber and Roehl, 1999)[16]. There have been other studies that have supported online experience or tenure as a key determinant of online buying behavior (Bellman, Lohse, and Johnson, 1999[18] and Beldona et al, 2004)[19]. Findings indicate that the greater the number of years the user spent online combined with higher frequency of Internet usage; the greater was the likelihood of buying (Bellman et al., 1999[18]; Weber and Roehl, 1999[16] and Beldona et al 2004)[19]. Alwitt and Hamer (2000)[20] posit that consumers increase their control with more time spent on the Internet, and in turn develop finer expectations of their interactions with businesses in general. Hammond, McWilliam, and Diaz (1998)[21] showed prior experience is an important moderator of users' attitudes towards the Web.

Based from the literature review of the Internet users, it is noticeable that Internet users are better educated, higher incomes, higher status occupations, have more online experience, use commercial lodging accommodations while travelling and tend to travel by air. In Malaysia, one of the profiles that had an early exposure to the culture of using the Internet for personal and professional reasons was university residents. These may consist of students, administrators and lecturers. However, university staffs are more likely to fit the profiles of the Internet users (in terms of purchasing) due to their financial capability and mobility. Consequently, this leads to:

H1: There is a relationship between the university staffs' demographic factors and the determinants of online hotel reservation

H2: There is a relationship between the university staffs' Internet usage behavior and the determinants of online hotel reservation

2.2 Determinants of online hotel reservation

There are various determinants identified in hotel reservations. The determinants of online hotel reservation intention used in this study are based from Salisbury et. al study (2001)[22]. It is based on the notion that perception is an i mportant determinant of individual behavior and that it, in turn, enhances the intention to book online. In this study, both perceived utility and risk of booking online is evaluated to examine the reasons why consumers do it.

Convenience, price comparison, and lower prices were identified as the three main reasons why Internet users buy travel products online (Starkov & Price, 2003)[23]. Because online consumers enjoy convenience and more control through making transactions online, online travel agencies should compete for customers through enhanced websites equipped with more convenient and ease of use features. Price is stated as the key motivator for consumers to purchase online (Santoma and O'Connor, 2006)[24]. Low prices has been found to be a major driver of online travel purchasing (Starkov & Price, 2003[23]; PhocusWright Report, 2000)[25]. According to the 2001 Annual PhoCusWright Travel Consumer Trends Survey (PhoCusWright, 2001)[26], almost 60% of online consumers cited price as the top reason to purchase travel-related products online and 53% of respondents believed that they get their best deals through online. Haussman (2002)[27] elaborated that online travelers are much more likely to make a reservation if they are offered the lowest price. Emmer et al. (1993)[28] stated that most of Internet users are searching over the web checking the price of tourism products before taking purchasing decision. Hence, online travelers would normally surf multiple web sites to research and compare prices before making their bookings.

Meanwhile, Szymanski and Hise (2000)[29] elaborated that consumer's perceptions of convenience; product information, site design, and financial security were the dominant factors in consumer assessments of online satisfaction. Technowledge's, 1999[30] findings stated that the top reasons given for shopping online were convenience, unique merchandise and competitive prices. According to CATALOG AGE's Consumer Shopping Survey (Chiger, 2001)[31], about 67 percent of the online shoppers agreed that convenience was the main reason they had bought via the Internet and 41 percent mentioned price as another factor. Hence, no wonder Swaminthan et. al (1999)[32] stated that the most compelling motivation to shop online is convenience.

Another factor that influences consumers to do online reservation is the flexibility and user friendliness of the system. With traditional reservation method, customers can easily change their particulars while talking to the hotel staff. They can effortlessly make any changes. However, when customers utilize hotel online reservation, such alterations are entirely the customers' responsibility. In the case of airline online reservation, any changes to prior reservations will be charged and there will be adjustments to the flight tickets in addition to availability. Consumers concern in this matter might have a "halo effect" on hotel online reservation.

Meanwhile, Jeong and Lambert's (2001)[33] empirical results showed that consumers' perceived quality of information about products and services on the web was most crucial in predicting their decisionmaking. Kim and Kim (2004)[34] study showed that online customers considered ease of information search and transaction to be more important factors than price. Although their studies provided evidence that convenience, price, and safety were significant factors for both online and non-online customers, their findings demonstrate the importance of the quality of information on the travel websites. Kim and Kim (2004)[34] findings corroborated Chu (2001)[35] and Zickefoose (2001)[36] studies where both studies indicated the importance of information. Shim et al. (2001)[17] investigated the relationship between the consumer's intention to use the Internet to search for useful information and an intention to purchase. The results of their study revealed that the intention to search the Internet for useful product information is the key predictor of the consumers' intent to purchase. In other words, providing useful and relevant information directly on the web is an important feature that affects online consumers' intention to purchase.

Finally, trust or security is the main concern for many consumers, particularly when planning leisure travel (Lewis and Semejn, 1998)[37]. Lack of trust is one of the most frequently cited reasons for consumers not shopping on the Internet (Turban et. al, 2002)[38]. Security is one of the most challenging and critical issues facing online system today. According to the survey by Tyler Nelson Software Interactive (2002)[39], security-related problems are the major barriers that prevent customers from purchasing online. The results showed that 30% of respondents who have not purchased online indicated problems regarding credit cards and 28% of those indicated general security concerns as reasons for avoiding online purchase. Yang and Jun (2002)[40] argued that both Internet purchasers and Internet non-purchasers consider security as their most critical concern. For those who decline to purchase online, the most significant reason was their fear of security. Shim et al. (2001)[17] and Jeong and Lambert (2001)[33] also supported the idea that online retailers need to build secure websites since Internet users hesitate to purchase products or services online due to security concerns. Thus:

H3: There is a relationship between the determinants of online hotel reservation and the intention to reserve online.

2.3 Behaviour Intentions

Behavioural intentions has been defined by Eagly and Chaiken (1993)[41] as a person's conscious plan to exert effort to carry out a particular behavior with these intentions being formed from both a personal evaluative and normative construct. In the seminal work of Zeithaml et al. (1996)[42], the behavioural intentions of customers are conceived as either favourable or unfavourable. Consequently, Malholtra and McCort (2001)[43] argue that generating a greater understanding of consumers' behavioural intentions continues to be a primary concern for marketing researchers. This is reflected in the frequency and rigor with which researchers have explored and modeled the antecedents of the behavioural intentions of offline consumers (e.g. Parasuraman and Grewal, 2000)[44]. However, a number of commentators have argued that more research is required to elucidate the nature and dynamics of online behavioural intentions (Chang et al., 2004)[45]

Based from the discussions and elaborations of the demographic characteristics, Internet usage behavior and the factors that influence the intention of online hotel reservation, a theoretical framework is developed and illustrated in Figure 1.

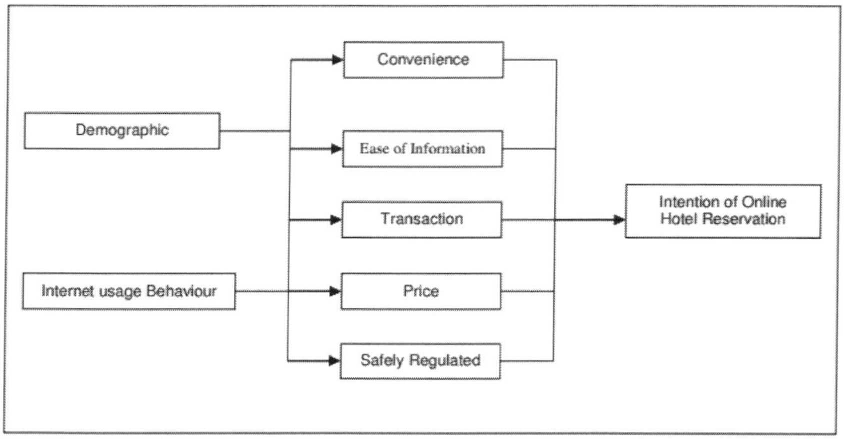

Figure 1. Research Theoretical Framework

3. METHODOLOGY

The respondents in the study are Universiti Utara Malaysia faculty members. A significant market segment of the Malaysian hotel industry comprises of government sectors. These consist of all employees working under the government linked institutions, ministries and state administrations. Government employees are entitled for special hotel rates which are about twenty per cent lower than the listed price. Universiti Utara Malaysia being a public university located at the very north of the country entails its staff to frequently travel outside the state for various official purposes such as meetings, research activities and etc. In addition, the university has thirteen centres through the country where lecturers would frequently travel and board a hotel room throughout the duration. Therefore, the selection of the university's faculty members could provide an insight towards the online behaviour intention of hotel reservation.

There are 1345 staffs comprising of administrative and academic staffs from three main colleges which basically encompass thirteen faculties. The sampling frame was based from the university's e-mail list server directory. A bilingual questionnaire of English and Bahasa Malaysia (the country's national language) was constructed based on previous research. Data were collected combining the web and email. The Universal Resource Locater (URL) of the questionnaire web site was given to respondents through each faculty list server. In this process, fifteen (15) e-mails bounced due to e-mail space quota excess. The respondent's were given one week to answer the questionnaire after which the URL to the questionnaire was block (expired). Respondents submitted their answers online and a data file was automatically generated. No incentives were used in this study.

The interactive method had numerous advantages compared to mail survey. First, it is less expensive because no paper or postage was involved. Data collection time was shorter and time was saved in data entry because the data file was automatically generated. This method also enables the researcher to communicate with respondents by e-mail if they had concerns. The questionnaire consisted of four sections. The first and second sections were designed to obtain the respondents demographic and Internet usage behaviour such as gender, age, income, education level, browser use per week, and number of years of Internet use and past online purchase experiences (1=yes, 2=no). The third section seeks customers' reasons of using the Internet for (hotel) purchase decisions. Fifteen (15) attributes were designed to measure the online hotel customers' perceived importance of each determinant. The determinants were developed from previous studies by Jarvenpaa and Todd (1997)[46], Weber and Roehl (1999)[16] and Jeong and Lambert (2001)[33]. The respondents were asked to rate each determinant using a 5-point likert scale (1=strongly disagree, 3=neither agree nor disagree; 5=strongly disagree). The final section measured respondents' online hotel reservation intention using a similar Likert style scale, purpose of reservation and the choice of the hotel standards.

194 responses were received where one case has to be deleted due to large sections of missing data. This gives a clean response rate of 14.51 per cent. The collected data were analyzed using Statistical Package for Social Sciences (SPSS) version 14.0. Statistical techniques such as descriptive statistics, factor analysis, and multiple regression analysis were used to achieve the objectives of this study. First, simple frequencies were generated to display the

distribution of respondents' demographic profiles and their Internet usage behaviour. Second, exploratory factor analysis (EFA) with varimax rotation was employed to determine the determinants of using the Internet for (hotel) purchase decisions. The EFA analysis verified the five factors as predetermined in the questionnaire development. However, some of the items were deleted or were move to represent other determinant. Finally, correlation and regression analyses were conducted to investigate the impact of the extracted factors and the likelihood of the respondent to do online hotel reservation.

Table 1. Factor Analyses, Mean and Cronbach's α

Item	Factors	Mean	SD	Percentage of Variance	Loading	α value
Factor 1	**Convenience**	4.32	0.68	18.52		0.86
	Easy to use	4.34	0.80		0.615	
	Convenient	4.39	0.74		0.591	
	24 hour accessibility	4.49	0.74		0.762	
	Variety of choice	4.07	0.95		0.813	
Factor 2	**Price**	3.53	0.88	17.70		0.83
	Better pricing compared to manual reservation	3.38	1.05		0.862	
	Reduced purchase related costs	3.79	1.00		0.695	
	Discounted price	3.40	0.99		0.866	
Factor 3	**Transaction**	3.55	0.81	13.73		0.72
	Trust in online transaction	3.28	0.97		0.681	
	Fast and efficient transaction	3.82	0.86		0.737	
Factor 4	**Easy and User Friendly**	4.08	0.76	13.00		0.81
	Ease of comparing hotels/services	4.07	0.96		0.772	
	Ease of placing order	3.96	0.90		0.763	
Factor 5	**Safely regulated**	2.94	0.81	12.57		0.76
	Security of sensitive information	3.01	0.98		0.836	
	Protected by law	2.88	0.84		0.835	

Tables 1 demonstrate the results of factor analysis, reliability scores of the construct, the mean and standard deviation data of the determinants of online hotel reservation. As can be observed in table 1, the most critical attribute was found to be convenience (mean=4.32) followed by easy and user friendly (mean=4.08), transaction (mean=3.55), price (mean=3.53) and safely regulated being the least importance (mean=2.94). Additionally, to examine the reliability of the determinants of online hotel reservation, a reliability test (Cronbach's α) was undertaken. The result revealed that the scales had scores ranging from 0.72 to 0.86 and were found to be reliable (Hair et al., 1998)[47].

4. FINDINGS

Table 2 shows the demographic and the Internet usage behavior of the respondents. As shown in Table 2, over half of the respondents are in the age of 30 to 40 years old with a majority of them are married. Based on the education level and salary, over 80% of the respondents seem to represent the profiles of an academician. A majority of the respondents access Internet at work and have been using it for more than 5 years. In terms of weekly usage, it varies almost evenly while 80 per cent of the respondents have made previous purchase from the Internet.

Correlation analyses were first employed to examine the relationship between demographic characteristics and Internet usage behavior on the factors that influence online hotel reservation. The results in Table 3 indicated that there's a significant positive ($t= 0.606, p<0.001$).

Finally, a regression analyses were employed to examine the effects of the determining factors on the likelihood of using online hotel reservations. The results in Table 4 demonstrate that transaction has the greatest effect followed by price. The next factor was ease of information and convenience being the last factor. Meanwhile, the findings for safely regulated did not show any significant relationship.

5. DISCUSSIONS AND CONCLUSION

The findings of the study reveal some interesting facts. Based from the demographics and Internet usage behaviour, heavy Internet users requires fast and efficient transaction. These findings are supported by the scores where the respondents with highest education, salary and number of years using the Internet have a significant relationship with previous experience of purchasing through the Internet. It can be implied from these scores that the respondent with the highest education are exposed to a longer tenure of Internet experience. These characteristics resembled academicians which comprised the majority of staffs in the faculty.

Table 2. The Respondents Profile

	Items	Details	Frequency	Percent
1	Age	20-30	46	23.8
		30-40	109	56.48
		40-50	31	16.06
		More than 50	7	3.66
2	Gender	Male	79	41.36
		Female	112	58.64
3	Marital status	Single	29	15.10
		Married	159	82.81
		Divorced	4	2.09
4	Highest Education	SPM and below	6	3.6
		STPM or Diploma	7	3.6
		Degree or Professional	16	8.3
		Postgraduate	164	84.50
5	Gross salary per month	Less than RM1500	6	10.89
		RM1501 - RM2500	15	7.77
		RM2501 - RM3500	50	25.90
		RM3501 - RM4500	69	35.75
		More than RM4500	53	27.46
6	Internet Access	At work	165	85.5
		At home	27	14.00
		Mobile access	1	0.50
7	No of years using the Internet	1 - 2 years	4	2.09
		3 -4 years	8	4.16
		More than 5 years	180	93.75
8	Usage of Internet on weekly basis	Less than 7 hours	26	13.54
		7 - 14 hours	48	25.00
		14 - 21 hours	46	24.00
		More than 21 hours	72	37.46
9	Previous purchase from the Internet	Yes	153	80.10
		No	38	19.9

Table 3. Correlation Scores

Factor	Age	Education	Salary	Years using Internet	Weekly Internet usage	Purchase from the Internet	Convenience	Information	Transaction	Price	Safety
ge	1										
ducation	.103	1									
Salary	.481**	.606**	1								
ears using nternet	-.002	.272**	.166*	1							
Veekly nternet sage	-.213**	.133	.078	.150*	1						
urchase rom the nternet	-.166*	-.361**	-.389**	-.306**	-.051	1					
onvenience	-.054	.148*	.043	.020	.095	-.114	1				
nformation	-.029	.070	.026	.051	.106	-.106	.652**	1			
ransaction	-.034	.051	.112	.051	.171*	-.199**	.612**	.559**	1		
rice	-.021	.076	.084	-.036	.117	-.067	.414**	.441**	.381**	1	
afety	-.032	-.085	-.046	-.106	.070	.012	.271**	.389**	.505**	.398**	1

Table 4. Regression Analyses of the Determinants with Online Hotel Reservations

Factor	Std β	*t*	F	Adjusted R^2
Convenience	0.261	3.735***	13.95	0.063
Ease of Information	0.280	3.110**	12.13	0.104
Transaction	0.369	4.308***	15.02	0.180
Price	0.293	4.091***	16.38	0.243
Safely Regulated	0.099	1.291	13.49	0.245

**p<0.05 and ***p<0.01

With regards to the relationship of the "antecedents' with the determinants of online hotel reservation, the correlation scores depicted that "academicians" look for convenience. The results corroborate the means results of the determinants where convenience is considered as the most important attributes in the intention to use online hotel reservation. Their educational background and tenure of using the Internet seems to alleviate their concerns of security. Apart from that, their position as a government staffs enables them to get a special discount (government rate) which is usually cheaper than the conventional promotional rates. Hence, pricing may not be an issue to them since they are usually entitled to a certain price bracket. The means scores also indicated that the motivating factor for using an online hotel reservation is to reduce the purchase related costs. In addition, the academicians used the system to compare hotels and service within their "eligible" rates that are allocated for them.

This study empirically investigated the relationship of the demographic characteristics and Internet usage behaviour on the determinants of online hotel reservations. The results showed that respondents with higher level of education exerts more importance on the convenience factor while the usage level of Internet and prior Internet shopping experience influence the transaction factor. The findings are consistent with previous research (e.g. Weber and Roehl, 1999)[16] that demonstrate busy travellers (as proxy by education level) look for convenience. However, the correlation scores depicted an interesting finding. Previous experience purchasing through the Internet has a negative correlation score with transaction. This illustrates that the respondents are unsatisfied with their initial experience purchasing through the Internet. Therefore, hotels need to improve their online reservation system as the online bookers look for reliable, efficient and fast online transactions.

In terms of the determinants on the likelihood of online reservations, the results suggested that online bookers place great importance on transaction factors which are the ability to make fast and efficient online transactions. Underlying the determinant factors is price. Online bookers used online hotel reservation to reduce their purchase related costs and simultaneously get good rates as well. Price is again the key motivator for consumers to purchase online as showed in previous studies (Santoma and O'Connor, 2006)[24]. Following this, online consumers expect the system to be user-friendly where one could compare hotels and is accessible 24 hour. The non significant safely regulated factor illustrated that online bookers are not aware of the regulations that govern or provide protections in the online commercial activities. The low scores demonstrate that online transactions are still view as not being well protected as traditional business transactions.

In a nutshell, this study has identified the target market profiles of “government sectors” online bookers. Based on the study findings, if Malaysian hotels were to set up their online booking system, they need to tailor their segmentation and positioning strategies to this group in addition to the various promotional activities such as introducing packaged products, sending electronic newsletters, and sending catalogs and direct mail to appropriate market segments. In ensuring the success of the online booking system, special emphasis needs to be focus on the transactional abilities, convenience and ease of information. Attractive pricing strategies need to be carefully crafted to lure online bookers to make reservations.

Finally, the findings of this study should be useful to online travel planners and marketing professionals in the travel and tourism industry in formulating their marketing strategies to fully exploit the potential of this new channel. Travel industry, travel planners and marketers are advised to take this market segment and determinants into consideration as they develop their online travel products.

6. LIMITATIONS OF STUDY

Although the study conforms to previous findings of online hotel reservations, the study's findings are limited by the sample (of a local public university). Great care is needed in applying the findings of the research to other populations. The response rate of 14.51 per cent was acceptable for this study because the sample was homogenous. For other general population samples, a higher response rate would be desirable.

7. RECOMMENDATIONS FOR FUTURE RESEARCH

A replication of this study is recommended using the general population. This will provide a test of reliability of this study's findings. Further research is also needed to differentiate the determinants of repeat online hotel bookers as compared to first time bookers. These will help hoteliers to differentiate the concerns of first time bookers as compared to season travelers that might request for special needs. The findings would definitely help hoteliers to more effectively retain these online hotel bookers.

REFERENCES

1. Eric, N. and Cassidy, F. Brown, L.,"Exploring The Major factors Influencing Consumer Selection of Travel Agencies in a Regional Setting" Journal of Hospitality And Tourism Management (13:1), 2006, pp26-42.

2. Connolly, D.J., Olsen, M.D., Moore, R.J., "The Internet as a Distribution Channel", Cornell Hotel and Restaurant Administration Quarterly, August, pp. 43-54.

3. Starkov, M., & Price, J., Online Travelers Prefer Booking Directly On The Hotel Website Retrieved April 15, 2003 from http://www.wiredhotelier.com/news/4015607.html.

4. Malaysia Internet Usage Stats and Marketing Report. Retrieved April 1, 2008, from http://www.internetworldstats.com/asia/my.htm

5. Cline, R.S., Warner, M., Hospitality e-business: the Future. Retrieved July 16, 2001 from http://www.hftp.org/Members/ BottomLine/ current/e business.htm.

6. Cobanoglu, C., Analysis of Business Travelers' Hotel Selection and Satisfaction,Ph.D. Dissertation,Oklahoma State University.

7. Peterson, R.A., Balasubramanian, S., Bronnenberg, B.J., "Exploring The Implications of The Internet for Consumer Marketing", Journal of the Academy of Marketing Science, (24), 1997, pp. 329-46.

8. Meuter, M.L., Ostrom, A.L., Bitner, M,J. and Roundtree, R., "The Influence of Technology Anxiety on Consumer Use and Experiences with Self Service Technologies", Journal of Business Research, (56:11), 2003, pp. 899-907.

9. Pernsteiner, C. and Rauseo, N., "Transforming the Hospitality Industry into e-Business", FIU Hospitality Review, (18:2), 2000, pp.10-21.

10. Grewal, D. Lindsey-Mullikin, H., "Loyalty in eTailing: A Conceptual Framework", Journal of Relationship Marketing, (2:¾), 2004, pp. 31-49.

11. Zeithaml, V.A., Parasuraman, A. and Malhotra, A., "Service Quality Delivery Through Websites: A Critical Review of Extant Knowledge", Journal of the Academy of Marketing Science, (30:4), 2002, pp. 362-375.

12. Goode, M. And Harris, L., "Online Behavioural Intentions: An Empirical Investigation of Antecedents and Moderators", European Journal of Marketing, (41: 5/6), 2007, pp. 512-536.

13. Jones, M.A., Mothersbaugh, D.L. and Betty, S.E., "Switching Barriers and Repurchase Intentions in Service", Journal of Retailing, (76:2), 2000, pp. 259-274.

14. Wu, Zhu, & Xu, "Trends in China's domestic tourism development at the turn of the century", International Journal of Contemporary Hospitality Management, (12:5), 2000, pp. 296–299.

15. Bonn, M.A., Furr, H.L. and Susskind, A.M. "Using The Internet As A Pleasure Travel Planning Tool: An Examination of The Sociodemographic and Behavioral Characteristics Among Internet Users And Nonusers", Journal of Hospitality and Tourism Research, (22:3), 1998, pp. 303–317.

16. Weber, K. and Roehl, W.S., "Profiling People Searching For And Purchasing Travel Products On The World Wide Web", Journal of Travel Research, (37:3), 1999, pp. 291–298.

17. Shim, S., Eastlick, M.A., Lotz, S.L., Warrington, P., "An Online Prepurchase Intentions Model: The Role of Intention to Search", Journal of Retailing, (77:3), 2001, pp. 397-416.

18. Bellman, S., Lohse, G. and Johnson, E.J., "Predictors of Online Buying Behavior", Communications of the AC, (42:12), 1999, pp. 32– 38.

19. Beldona, S., Kline, S., & Morrison, A. M., "Utilitarian Value In The Internet: Differences Between Broadband & Narrowband Users", Journal of Travel & Tourism Marketing, 2004, pp. 17-23.

20. Alwitt, L. F., & Hamer, L. O., "The Effect Of Internet Experience On Consumer Expectations Of Responsiveness And Control In Offline Services Marketing Interactions", Proceedings of the 33rd Hawaii international conference on system sciences, January 4–7, 2000, Maui, Hawaii.

21. Hammond, K., McWilliam, G. and Diaz, A.N. "Fun And Work On The Web: Differences In Attitudes Between Novices And Experienced Users", Advances in Consumer Research, (25), 1998, pp. 372-378.

22. Salisbury, W.D., Rodney, A.P., Allison, W.P., Miller, D.W., "Perceived Security and World Wide Web Purchase Intention", Industrial Management and Data Systems, (101:4), 2001, pp. 165-177.

23. Starkov, M, Price, J. Hoteliers' 2007 Top Ten Internet Marketing Resolution, Retrieved September 1, 2007, from www. eyefortravel.com/index.asp?news=54142.

24. Santoma, R., O'Connor, P., The Online Pricing Practices Of Up-Market Barcelona Hotels: An International Comparison, in Hitz, M., Sigala, M., Murphy, J. (Eds). Information and Communication Technologies in Tourist. SpringerWien, New York, NY, 2006, pp. 221-33.

25. PhocusWright Report. Consumer Behavior, Patterns, Attitude and Outlook (Online) Travel ECommerce Survey. Retrieved April 19, 2002, from http://www.phocuswright.com .

26. PhocusWright Report. Consumer Behavior, Patterns, Attitude and Outlook (Online) Travel ECommerce Survey. Retrieved May 1, 2003, from http://www.phocuswright.com .

27. Haussman, G. WorldRes survey warns of industry commoditization. Retrieved February 2,2002.From http://www.hotelinteractive.com/ news/articleview.as p? articleID=1392.

28. Emmer, R.M., Tauck, C., Wilkinson, S., Moore, R.G., "Marketing Hotels Using Global Distribution System", Cornell Hotel and Restaurant Administration Quarterly, (34:6), 1993, pp. 80-89.

29. Szymanski, D.M., Hise, R.T., "E-Satisfaction: An Initial Examination", Journal of Retailing, (76:3), 2000, pp. 309–322.

30. Kau, A.K., Tang, Y.E., Ghose, S., "Typology of Online Shoppers", Journal of Consumer Marketing, (20:2), 2003, pp. 139-156.

31. Chiger, S. "Consumer Shopping Survey", Catalog Age, November, 2001, pp. 80-81.

32. Swaminathan, V., Lepkowska-White, E. and Rao, B.P., "Browsers Or Buyers In Cyberspace? An Investigation Of Factors Influencing Electronic Exchange", Journal of Computer Mediated Communication, (5:2), 1999, pp. 1–23.

33. Jeong, M., Lambert, C.U., "Adaptation of an Information Quality Framework to Measure Customers' Behavioral Intentions To Use Lodging Web Sites", International Journal of Hospitality Management, (20:2), 2001, pp. 129–146.

34. Kim, W.G. and Kim, D.J., "Factors Affecting Online Hotel Reservation Intention Between Online and Non-Online Customers", International Journal of Hospitality Management, (23:4), 2004, pp. 381- 395.

35. Chu, R., "What online Hong Kong travelers look for on airline/travel websites?", International Journal of Hospitality Management, (20:1), 2001, pp. 95–100.

36. Zickefoose, J. (Ed.), "Websites 101", Lodging Hospitality, (57:8), 2001, pp. 22–24.

37. Lewis, I. and Semeijn, J., "The Impact of Information Technology on Travel Agents", Transportation Journal, (37:4), 1998, pp. 20-6.

38. Turban, E., King, D., Lee, J., Warkentin, M. and Chung, H., Electronic Commerce: A Managerial Perspective, 2002, USA Prentice Hall.

39. Tyler Nelson Software Interactive, Global ecommerce report 2002. Retrieved 27 August 2002 from http://www.tnsofres.com/ger2002/home.cfm

40. Yang, Z. and Jun, M., "Consumer Perception Of E-Service Quality: From Internet Purchaser And Non-Purchaser Perspectives", Journal of Business Strategies, (19:1), 2002, pp. 19–41.

41. Eagly, A. and Chaiken, S. The Psycology of Attitudes, 1993, Harcourt Brace Jovanovich, TX.

42. Zeithmaml, V.A., Berry, L.L. and Parasuraman, A., "The Behavioural Consequences of Service Quality", Journal of Marketing, (60: April), 1996, pp. 31-46.

43. Malholtra, N.K. and McCort, J.D., "A Cross Cultural Comparison of Behavioural Intentions Models", International Marketing Review, (18:3), 2001, pp. 235-269.

44. Parasuraman, A. and Grewal, D., "The Impact of Technology on Quality-Value-Loyalty Chain: A Research Agenda", Journal of the academy of Marketing Science, (28:1), 2000, pp.168-174.

45. Chang, L., Marchewka, J.T. and Ku, C., "American and Taiwanese Perceptions Concerning Privacy, Trust and Behavioural Intentions in Electronic Commerce", Journal of Global Information Management, (12:1), 2004, pp. 18-40.

46. Jarvenpaa, S.L., Todd, P.A., "Consumer Reactions To Electronic Shopping On The World Wide Web", International Journal of Electronic Commerce, (1:2), 1997, pp. 59–88.

47. Hair, J.F., Anderson, R.E., Tatham, R.L. and Black, W.C., Multivariate Data Analysis (5th ed.), 1998, Prentice-Hall International, London.

14

RESEARCH ON THE APPLICATION OF FAULT TREE ANALYSIS FOR BUILDING FIRE SAFETY OF HOTELS

Yi-nan Hu

Chinese People's Armed Police Froces Academy , Langfang 065000, Hebei Province, China

ABSTRACT

Nowadays, in order to prevent large-scale casualties and significant property losses in the society, strengthening the building fire safety of hotels has become one of the hottest topics explored by more and more experts and scholars. In accordance with the building fire system analysis of hotels, considering the whole process of the fire accident, the fault tree of building fire was established. The analysis of relationship between the primary and secondary factors was conducted from three parts, the minimal cut set, the minimal radius set and the importance of structure. There are three main reasons for building fires in hotels, involving the personnel's unsafe behavior, the failures of electrical equipment and wiring as well as the ignition of flammable and combustible substance. And the key measures for prevention of building fires including strengthening the management of daily fire safety, key positions and key personnel have been proposed, which can improve the safety and operational reliability of hotel buildings. The analysis shows that the fault tree analysis method can make a comprehensive exposition of various factors causing fire of hotel buildings and their logical relations, which also provides a reference for the improvement of the design and the construction of the safety technical measures in other fields.

KEYWORDS

Hotel; Building Fire; Fault Tree; Importance of Structure

1. INTRODUCTION

In recent years, building fires of hotel occur frequently, resulting in heavy casualties and serious economic losses. On June 10, 2005, the fire of Huanan hotel in Shantou caused a total of 31 deaths and 28 people injured, with the direct property losses of 810,586 Yuan. It was an electric short circuit on the ceiling in front of the door of a private room on the second floor of the building lead to this fire, which ignited surrounding combustible materials. On April 14, 2013, the fire of Xiangyang garden hotel was caused by another electrical short circuit in the internet bar on the second floor, leaving a total of 61 casualties, including 14 people killed. On May 1, 2011, there was another building fire in an express hotel in Tonghua because of the arson in the bar on the first floor of the underground, resulting in 10 people killed, 2 people seriously injured and 51 people slightly injured.

A modern hotel generally has the characteristics of being multi-functional and comprehensive, since it is a place to provide passengers with accommodation, dining, entertainment, meeting and banquet. As a concentrated "small society", there is a great hazard given by hotel building fires. The causes for fires in hotel include the complexity of the function, the luxury of the decoration, varies of internal fuel as well as the frequent use of electricity. In addition, the ideological paralysis of staff and the guests' weak consciousness of safety increase the possibility of building fires. In the event of fire, flames and high temperature flue gas will spread quickly, which can result in serious accidents with heavy casualties and property damages. Therefore, to prevent the building fires of hotels has become one of the issues to be solved urgently in the field of fire safety

As an effective method for the reliability analysis and evaluation of the safety system engineering, the fault tree analysis method provides an effective means for the analysis of fire accidents in hotel buildings. At the same time, it also has a good social benefit and a practical prospect of application in various fields of the safety management. With the help of analysis of the fire process in hotel buildings, the fire mechanism and causes can be analyzed gradually, and the key factors of building fire prevention can be found out. Therefore, taking some effective safety measures will minimize the incidence of malignant casualties, which also provides scientific and credible reference for safety assessment.

2. BASIC THEORY OF FAULT TREE ANALYSIS METHOD

Fault Tree Analysis∧FTA∨, as a method for system reliability analysis, is considered as one of the simplest, most effective and prospective tools for analyzing some complicated systems. By applying this method, a logical relationship of the system between the accident that might occur and the cause of the accident in the form of the arborescence will be built. The main cause of the accident can be found out through the qualitative and quantitative analysis of the fault tree. It can provide a reliable basis for the determination of some safety countermeasures, in order to achieve the purpose of prediction and prevention of the accidents. Currently, it is a more comprehensive and practical technology method of the accident prevention and risk analysis, widely adopted in the identification, prediction, evaluation and control of accidents.

Fault tree analysis program usually changes with the difference of the object of evaluation, the purpose of the analysis and the degree of thickness. But it generally works in accordance with the following procedures, as shown in figure 1.

3. CONSTRUCTION OF FAULT TREE OF BUILDING FIRES OF HOTELS

3.1. Analysis of building fire causes of hotels

According to the statistic data of 294 building fires in hotels in the past ten years, from 2006 to 2015, occurred in China, the main causes for fire have been known such as staff negligence, careless construction, equipment and facilities malfunction, electrical wiring short circuit, improper operation as well as artificial arson. The personnel's unsafe behavior, the failures of electrical equipment and wiring as well as the ignition of flammable and combustible substance are considered as main three reasons, and the rate of the hotel building fire accidents triggered by them is as high as 90.17%. The rate of the hotel fire accident caused by the staff's unsafe behavior is as high as 36.94%. There are various causes for fire including the staff's wrong operation, violation of relevant regulations in the process of work, and even staff's poor and unqualified skills for preventing fire. More seriously, some residents are short of consciousness for fire control safety. While smoking, they set a fire in the hotel room by lighting a fire on the carpet in the hotel. Due to improper installation of the electrical equipment and electrical wiring, the bad contact of wire, the poor heat dissipation, and an increase of electric circuit load led by large-scale electric equipment, the ratio of hotel fires caused by electrical accidents is 27.85%. Coupled with the long running of equipment, the rise of temperature and the damage on surface insulation of the wire, the leakage or short circuit will occur in case of lacking of the maintenance of electrical equipment and wiring, which can lead to building fires in hotel. The proportion of fires triggered by external source which ignites the flammable and combustible materials reaches 25.38%, and most of them are caused by the mars splashed during the constructions and repairs.

3.2. Fault tree graph of building fire in hotels

On the basis of the investigation of causes for large hotels, the fire in the hotels can be regarded as a top event. According to the analysis and study of the above process of fire accidents, every middle event and basic event for the fire can be determined .To show the interrelationship among top event, middle event and basic event, a fault tree of the fire in the hotel will be established, as shown in figure 2.

4. ANALYSIS OF FAULT TREE OF BUILDING FIRE IN HOTELS

4.1. Determination of minimal cut (radius) set

Based on the logic simplification approach and Boolean algebra simplifying method, we can deduce the minimal cut set of this fault tree is 23, while the minimal radius set is 4. The qualitative analysis is easy to start with the minimal radius set, and it is more convenient to choose the solutions to control the accidents.

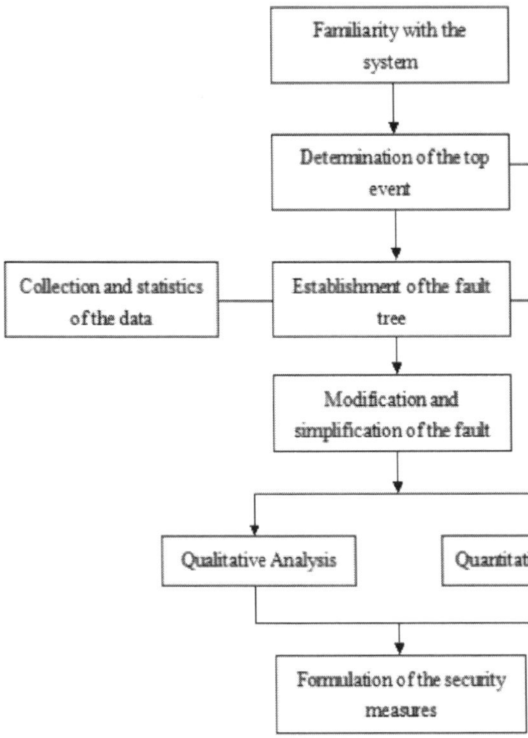

Figure 1. General procedure of fault tree analysis

A direct method to solve the minimal radius set is to use the duality of the source tree. According to Morgan's theorem, every accident should be transferred into antithetic events. Then the successful tree of this accident can be made. After this, Boolean Algebra is adopted to simplify the solution, so four minimal radius set can be got as follow:

$P_1=\{X_1,X_2,X_4,X_6,X_7,X_8,X_{12},X_{13},X_{14},X_{15},X_{16},X_{17},X_{18},X_{19},X_{20},X_{21},X_{22}\}$

$P_2=\{X_1,X_2,X_4,X_5,X_6,X_9,X_{10},X_{11},X_{12},X_{13},X_{14},X_{15},X_{16},X_{17},X_{18},X_{19},X_{20},X_{21},X_{22}\}$

$P_3=\{X_1,X_3,X_4,X_6,X_7,X_8,X_{12},X_{13},X_{14},X_{15},X_{16},X_{17},X_{18},X_{19},X_{20},X_{21},X_{22}\}$

$P_4=\{X_1,X_3,X_4,X_5,X_6,X_9,X_{10},X_{11},X_{12},X_{13},X_{14},X_{15},X_{16},X_{17},X_{18},X_{19},X_{20},X_{21},X_{22}\}$

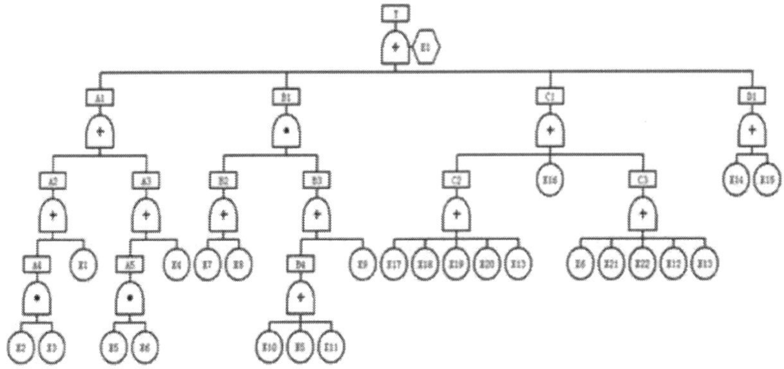

Figure 2. Fault tree graph of building fire in hotel

Table 1. Symbols and meanings of fire fault tree in large hotels

Code	Basic Event	Code	Basic Event
T	Building fire in hotels	X5	Smoking of hotel guests
A1	Personnel's unsafe behavior	X6	Peripheral flammable and combustible substance
A2	Staff's unsafe behavior	X7	Use of flammable and combustible decoration materials
A3	Resident's unsafe behavior	X8	Other flammable and combustible substance
A4	Fire of the pipeline	X9	Electric spark
A5	Smoking	X10	Welded construction with fire
B1	Ignition of flammable and combustible substance	X11	Fireworks
B2	Flammable and combustible substance	X12	Insulation breaking
B3	Fire source	X13	Poor heat dissipation
B4	Open fire	X14	Arson
C1	Failures of electrical equipment and wiring	X15	Natural disaster
C2	Improper use	X16	Poor quality of the electric wire
C3	Improper installation	X17	Improper operation of electrical equipment
D1	Other fires	X18	Frequent power electricity
E1	Mismanagement of fire prevention	X19	No power off
X1	Illegal operation	X20	Using of electrical equipment overloaded
X2	Pipeline oil accumulation	X21	Unreasonable line installation
X3	Mars in the pipeline	X22	Line contact undesirable
X4	Illegal use of fire		

4.2. Determination of the importance of structure

Based on the fault tree structure, the importance of structure analyzes the importance degree of each basic event without consideration of the probability of occurrence of each event. Or it is assumed that the probability of occurrence of each event is equal. Then the influence degree of an occurrence of the basic event on the top event is analyzed.

Because different basic event has different impacts on the top event in the fault tree analysis, in order to understand and know well the influence degree of an occurrence of the basic event on the top event, the author adopts the minimal radius set and ranking method to work out an importance degree of structure:

$$I[X_1]=I[X_4]=I[X_6]=I[X_{14}]=I[X_{15}]=I[X_{16}]=I[X_{17}]=I[X_{18}]=I[X_{19}]=I[X_{20}]=I[X_{21}]$$

$$=I[X_{22}]=I[X_{12}]=I[X_{13}]>I[X_7]=I[X_8]>I[X_2]=I[X_3]>I[X_5]=I[X_9]=I[X_{10}]=I[X_{11}]$$

4.3. Result Analysis and Countermeasures

4.3.1. Result Analysis

(1) From the analysis of fault tree chart, there are 11 or-gates and 3 and-gates. It follows from the definition of or-gate that if any input event happens, it can lead the occurrence of output events. According to the proportion of or-gates and andgates, it is obvious that there is a high fire accident risk in the buildings like hotel.

(2) There are only 22 basic events in the fault tree with 23 minimal cut sets and 5 minimal radius sets. It shows that there are more channels of the occurrence of top events and less control channels, which proves that this system is relatively dangerous. Therein, the more the minimal cut sets are, the more dangerous the system is. So increasing the basic events of the minimal cut sets will help to improve the system security and decrease probability of the occurrence of contingency. The more the minimal radius sets are, the safer the system is. As to the minimal radius set of the few events with less administration programs, it is more economical and effective for events minimal radius set.

(3) From the importance degree of structure, the basic events, such as employees' negligence and carelessness, fault of construction

equipment, short circuit, welding improper operation and arson, have a large proportion in the importance degree structure, which are the most important factors to the top event. However, the electrical sparkle and other events like open flames fire have a small proportion in the importance degree of structure, which are less important to the top event. Above all, if the basic events that have large importance degree can be controlled, the probability of occurrence will be much lowered.

(4) It can be known from the definition of the minimal radius set that if any one of P1, P2, P3 and P4 can be avoided, the top event won't happen anymore. For in these four precaution channels there are all kinds of basic events with the highest importance of structure, and the majority of the other basic events can be controlled by human efforts, the solution with the best precaution actions should choose the minimal radius sets with the least basic events, which is the solution of P1 and P3. But taking the possibility, reliability and structural importance degree of the prevention into consideration, the fire accident prevention in hotel buildings should be based on all the solutions, namely solution P1, P2, P3 and P4 to make preventive measures and have well preparations for the basic events that can be controlled by human efforts

(5) Besides, E1, as an important condition for hotel fire accident, should be controlled strictly. The deep causes of many hotel fire accidents are the lack of fire management and the problems of violation of the fire-control regulations and unimplemented fire management mechanism. Therefore, only E1 is strictly controlled, can it play an important role to prevent other fire accidents.

4.3.2. Countermeasures

Based on the fault tree analysis method, various factors of the fire in hotel buildings and their logical relationships have been made for offering a comprehensive elaboration, and the important factors in the occurrence of fire have been obtained correspondingly. In order to achieve the effect of the fire prevention, the occurrence of each basic event should be controlled, especially the events having a large proportion in the structural importance degree, such as accumulation of piping grease in kitchen, fault operation of electric equipment and unreasonable installation of circuit line. From the above analysis on the key link of initiation fire, some relevant countermeasures are suggested to be made.

(1) To strengthen the daily management of the key parts, effectively preventing fire caused by equipment and circuit lines. The design and construction work of electrical wiring and electrical equipment should be in a strict conformity with relevant provisions such as "Regulations for Low-voltage Electrical Installation ", "Specifications for Electrical Fire Protection Design "and "Code for fire protection design of buildings", and the laying and installation of cable and wire should be guaranteed, so as to eliminate fire hazards of electrical wiring, equipment from the source. Furthermore, it is essential to focus on the fire inspection of the power distribution room, cable trench, large load electrical equipment and other situation, which is useful to avoid the risk of wire aging, leakage and short circuit.

(2) To strengthen fire safety management of key personnel and carry out fire safety publicity and education training. Hotel fire safety education should be included in the staff education training plan, and fire safety special courses should be carried out by schedule. The staff need to pass the examination before going to their posts, which enables them to improve fire safety awareness and learn more knowledge of fire control. The training includes fire laws and regulations, fire safety knowledge as well as knowledge and skills, and the training will focus on the staff's capacity of organizing and guiding the evacuation of guests, reporting fire alarm and putting out early stage fire.

(3) To strengthen the daily management of fire safety and clean fire hazards. A system for hotel fire safety responsibility should be established properly. In the first, a strict post responsibility system should be established to determine the first responsibility of fire safety. In addition, fire safety inspection system need to be built with organizing regular fire safety inspections and checking to detect fire hazards in time. The emergency plans of firefighting and evacuation should be developed which is useful for organizing regular fire drill installations. Last but not least, it is necessary to strengthen the key period of fire safety management, namely, the night patrol management during 22 at night to 6 in the morning to discover and eliminate fire hazards as early as a fire occurs.

(4) To strengthen the inspection and maintenance of firefighting equipment facilities. Hotel should ask staff to detect fire automatic fire alarm system and its linkage function test, in order to ensure the normal operation of fire facilities. Once the firefighting equipment fail to work correctly, give the maintenance to them for the first time, to guarantee the equipment work safely within a short time.

5. CONCLUSION

Through the analysis of system specific fault conditions, fault tree analysis method can help to identify inherent or potential risk factors within the system and determine the defect of system, providing a scientific basis for the improvement of the system security design, the development of security technology countermeasures, the formulation of safety management measures and accident analysis. More importantly, based on the fault tree analysis method, weak links for the fire accidents can be found out and the improvement scheme will be put forward, which can help us avoid the fire accident to a certain degree. There are specific conclusions as follows:

(1) According to the analysis of the building fire in hotels, it is obvious that this kind of fire accidents mainly results from three causes including the personnel's unsafe behavior, the failures of electrical equipment and wiring as well as the ignition of flammable and combustible substance. Therefore, for the prevention of fire accidents, it is significant to control each factor effectively and manage each link well.

(2) In order to minimize fire hazards, based on the fire causes, ways of the prevention of fire have been put forward that it is essential to strengthen the management of daily fire safety, key parts and key personnel. In addition, the regular inspection and maintenance of equipment and facilities can achieve normal starting of firefighting facilities in the event of fire, which it is helpful to prevent fire spread. Only to seize the key measures, can it improve the security of hotel buildings

(3) The application of fault tree analysis method can be adopted to analyze the building fire accidents not only in the hotel, but also in the other buildings such as shopping malls and residences, in order to formulate specific prevention measures as the site reference.

ACKNOWLEDGEMENTS

My deepest gratitude goes first and foremost to Professor Li-zhi Wu , my supervisor, for his constant encouragement and guidance. He has walked me through all the stages of the writing of this thesis. Without his consistent and illuminating instruction, this thesis could not have reached its present form.

Second, I would like to express my heartfelt gratitude to Professor Ya-di Li, who led me into the world of translation. I am also greatly indebted to the professors and teachers at the Department of Fire Safety Engineering: Professor Xiao-bin Li, Professor Si-cheng Li, who have instructed and helped me a lot in the past two years.

Last my thanks would go to my beloved family for their loving considerations and great confidence in me all through these years. I also owe my sincere gratitude to my friends and my fellow classmates who gave me their help and time in listening to me and helping me work out my problems during the difficult course of the thesis.

REFERENCES

1. Xian-zhi Lv, Jian-yun Ma, Kun Tao, Feng Ma, Shuo Chen, Establishing of Fire Safety Evaluation System For High-rise Hotel Using Fuzzy Comprehensive Evaluation Method, Journal of Safety Science and Technology7, pp. 138-143

2. Cai-xia Hao, Yan Xu, Sheng-wu Gong, 2012. Application of Fault Tree Analysis on Fire and Blast Accidents of LPG Tank, Journal of Safety Science and Technology8, pp.154-159.

3. Ai-you Wu, Cong-lu, Guo-qing Xiao, 2005. Application of Fault Tree Analysis in Casualty of Building Fire, China Safety Science Journal15, pp. 92-95.

4. Jie Yin, Xiang-min Zheng, Bin-bin Dong, 2015.Study on Features of Fire Accidents in Hotel Industry and Construction of the Prevention and Control Management System, Safety and Environmental Engineering22, pp. 102-107.

5. Xiao-ping Zhang, 2013. On the Fire Safety of Restaurants and Hotels Based on Fire Cases, Journal of Chinese People's Armed Police for Force Academy29,pp.66-68.

6. Ying Zhang, 2013. Building Electrical Fire Accident Analysis and Safety Technical Measures, Scientific and Technology Innovation and Application6, pp. 236-243.

7. De-lu Liu, Zhi-qiang Wang, Yu Li, 2003. Introduction to the Current Situation of the Fire Facilities and Countermeasures of Hotels, Fire Science and Technology22, pp. 336-337.

INDEX